British Aircraft of WWI

Volume 3 | Fairey Aircraft

Colin A. Owers

Great War Aviation Centennial Series #57

Dedicated to Katrina whose courage in difficulties is an inspiration to me.

Acknowledgements

This book would not have been written without the help of many friends over many years. Ronny Bar; the late Jack M Bruce; Ian M Burns; Juanita Franzi; Jack Herris; Colin Huston; Sir Peter Jackson; the late G Stewart Leslie; Peter London; Byran Ribbans; the late Ian R Stair; Phil Tomaselli; and especially my wife Julie Owers who copied so many documents in various archives over the years. Any errors are mine alone.

For information on our aviation books, please see our website at: **www.aeronautbooks.com**. Aeronaut Books is looking for photographs of rare German aircraft of WWI for our books. To help please contact the publisher at **jherris@me.com**.

ISBN: 978-1-953201-16-4

Design and layout: Jack Herris
Cover design: Aaron Weaver
Cover Painting: Taras
Color Profiles: Ronny Bar
Digital photo editing: Jack Herris

Aeronaut Books

Books for Enthusiasts by Enthusiasts

www.aeronautbooks.com

Table of Contents

Chapter 1: Fairey Aviation 3

Chapter 2: The Fairey F.2 5

Chapter 3: The Fairey Campania 16

Chapter 4: Hamble Baby 63

Chapter 5: The Fairey N9 98

Chapter 6: The Fairey N10 and Type III to IIIC 112

Chapter 7: The Fairey N2a 196

Chapter 8:The Fairey N4 198

Chapter 1: Fairey Aviation

Charles Richard Fairey was born on 5 May 1887. His father suffered a business failure and then died with the result that the young Richard had to leave the Merchant Taylor's School at 15. He was employed at the Jandus Electric Co of Holloway that made arc lights. He studied at nights at the technical college and became an aviation enthusiast building elastic powered flying model aeroplanes. Through his model activities he met many pioneers of British aviation such as A.V. Roe. In his biography of Sir Richard Fairey, Adrian Smith remarks that Fairey's models of *minimum weight and maximum robustness* were harbingers of his later aircraft *of great delicateness yet remarkable tensile strength.*[1]

Fairey produced a model canard monoplane with twin opposite-rotating airscrews and an ingenious wing design to win the Kite and Model Aeroplane Association Challenge Cup in 1910. Fairey managed to convince the manager of A.W. Gamage & Co to purchase the rights to the model and with the money he received, said to be about £300, he formed a company to produce aircraft models and by November 1910 had supplied Gamage with more than 200 of them. Unbeknown to him, his model had infringed the patents of aviation pioneer John William Dunne. Dunne had been working for some time on his tailless swept-wing biplanes to achieve automatic stability. Instead of taking the young engineer to task, Dunne only required him to acknowledge the J.W. Dunne patent on the instruction booklet. Fairey so impressed Dunne that he took him into his company, the Blair Atholl Syndicate, at Eastchurch. He joined Dunne in 1911 giving up his steady position at the Finchley Power Station. It was here that he came under the influence of Naval aviators and the Short brothers who also occupied the aerodrome.

In late 1912, Fairey left Dunne and moved to Short Brothers, also at Eastchurch. His rise to Chief Engineer was proof of the trust that Horace Short placed in him. He considered Horace his mentor and the 30 months he spent with the company the best years of his life. He tried unsuccessfully to join the RNAS and RFC on the outbreak of war, but was rejected. The legend is that Cdre Murray F Sueter told him that he would make a better contribution to the war effort by building aeroplanes for the Navy. Fairey is reputed to have said he would do so if he could get a contract and start his own company, and so he formed the Fairey Aviation Co. With the approval of the Short brothers he gained a subcontract to build 12 Short 827 floatplanes.

Given facilities at the Army Motor Lorries Co (AMLC) at Clayton Rd, Hayes, where only about half the factory was required for the reconditioning operations of the lorry company, Fairey set up his first factory. A field at Harlington was bought and this was to be the future site of the Fairey factory. The assembly of the Short seaplanes took place in the shed erected there, however, in order to allow for the testing of seaplanes, the Admiralty made available a site at the mouth of the Hamble River, Southampton Water. Here the shops had to be built on concrete piers due to the marshy ground and it was here that Sidney Pickles, the company's first test pilot, test flew the 827 seaplanes after reassemble. These were followed

Below: A Fairey built Short 827. Of interest is the way the forward rudder stripe has caught the light to give a different tone to that painted on the fin. Trying to determine colours from black and white photographs is virtually impossible without other evidence.

Above: The Fairey-built 1½ Strutters were the two-seat fighter version of the design. Sopwith 1½ Strutter A993 of No. 43 Squadron, RFC, is shown in German hands after being forced down on 28 April 1917, at Roncy. The pilot was 2/Lt Cliff M Reece, with 2AM A Moult as observer/gunner who was wounded. The machine was credited to *OfStv* Edmund Nathanael of *Jasta* 5 for his 12th victory.

by another contract, this time for Sopwith 1½ Strutters. Fairey started work on his own designs, that became the F.1 and F.2. The F.2 was a twin-engine heavy fighter to accompany bombing raids on occupied Belgium. The machine that has become known as the F.2 is discussed in Chapter 2.

In order to gain more room Fairey apparently took over the rest of the Clayton Road works from AMLC by just moving into the works. Whether this story is true, Fairey did gain control and the order for Sopwith 1½ Strutters was completed in less than 12 months, 26 in 1916, the rest by September 1917.

Fairey's patent camber changing gear gave his company the contract to build a version of the Sopwith baby, the Fairey Hamble Baby. In the event, Parnall was to build more Hamble Baby types than Fairey, but the patent Camber Gear was to continue to bring dividends to the company in the post-war days when military aviation procurement was drastically reduced.

Due to the contract for the Hamble Baby the company took over the Admiralty's shed at Hamble, as their original site was inadequate. In 1917 a £20,000 loan from the Government enabled the erection of a plant at the Harlington site. The end of the war saw many aviation orientated companies burdened with these loans. Fairey was lucky that his IIIC was required urgently for the Allied intervention in Russia, and despite the sudden downturn in work as contracts were cancelled across the board, Fairey was able to keep working and keep his staff intact.

Post-war the Fairey III variants kept the company going when other faltered and disappeared altogether. The Fairey IIID was followed by the IIIF, cementing the relationship between Fairey and the Admiralty. For the story of Fairey post-WWI, Adrian Smith's biography (*The Man Who Built The Swordfish – The Life of Sir Richard Fairey,* I.B. Tarus, UK, 2018.) covers the dealing of the company with its acquisitions, sales, and Fairey's personal life, but less on the design and aircraft history part of the Fairey story.

Other Manufacturer's Aircraft Built by Fairey

Manufacturer	Type	Serial	Notes
Short Brothers	Admiralty 827	8550 to 8561	Deliveries commenced in March 1916.
Sopwith	1½ Strutter	A953 to A1053	Two-seat fighter

Chapter 2: The Fairey F.2

Above: The F.2 seen soon after assembly in the Harrington shop. The original five wooden sheds on the site are visible in the background. Note the tailplane support strut and lack of serial.

After Fairey had shown that they could build and deliver aircraft on time, they were given a contract to build two Air Department (A.D.) Heavy Bomb Dropper biplanes. Although not confirmed, this is thought to be the origin of the Fairey F.1. H.A. Taylor, in his *Fairey Aircraft since 1915,* (Putnam, UK, 1984), does not mention what aircraft received the designation F.1.

In December 1915, the Admiralty issued specifications to various firms inviting them to enter a competition to design two land machines, not seaplanes. The competition was to close in February. *The contractor will be given an absolutely free hand with regard to design and construction, and supervision will only be exercised with regard to the quality of material and workmanship employed. …*

If the machine proved to be up to Admiralty requirements it would be purchased at a price to be agreed upon and type trials would be carried out at which any alterations which were considered necessary would be made, a representative of the firm being present with the necessary draughtsmen to take records of all such alterations for future machines. The successful firm would receive an order for a certain number of their machines but would have to agree to have its design built by other firms either by sub-contract or by the Admiralty paying a royalty. Provision was to be made for dual control *and the machine should be inherently stable.*

The specifications for the machines were as follows:

(1). Bomb Dropping Machine.

Speed – not less than 85 land miles per hour at 7,000 feet when fully loaded.

Endurance – 450 miles in calm air plus the amount of fuel to reach 10,000feet, and one hour's additional fuel as a margin.

The machine is to carry at least one passenger in addition to the pilot, and at least one machine gun placed in such a position to effectively protect the machine from the attack of hostile aircraft over the maximum arc of vision possible. Weight of gun and ammunition was allowed for at 80 lbs. It was to carry 8 cwt of bombs.

The machine was to be capable of climbing to 12,000 feet in less than an hour. The Admiralty would provide a 240-hp Sunbeam engine for the machine.

(2). Fighting Machine.

Speed - Not less than 90 land miles per hour at 7,000 feet.

Endurance - 540 miles in calm air plus the amount of fuel to reach 10,000 feet. Fully loaded it was to be able to climb to 10,000 feet in not more than 40 minutes.

Useful Load – Was the pilot and two passengers, at least two machine guns. The weight of the machine guns and ammunition was to be taken as 100 lbs each. The ammunition was to be stowed *near the gunner's positions in suitable chargers. The machine guns must be placed in such positions a to obtain the maximum cones of fire for both guns, and these cones of fire for the two guns must overlap to the maximum extent.* It was also desired that the machine carry 200 lbs weight of bombs. *Should you, however, prefer to put in the equivalent weight of armour so as to protect the pilot and*

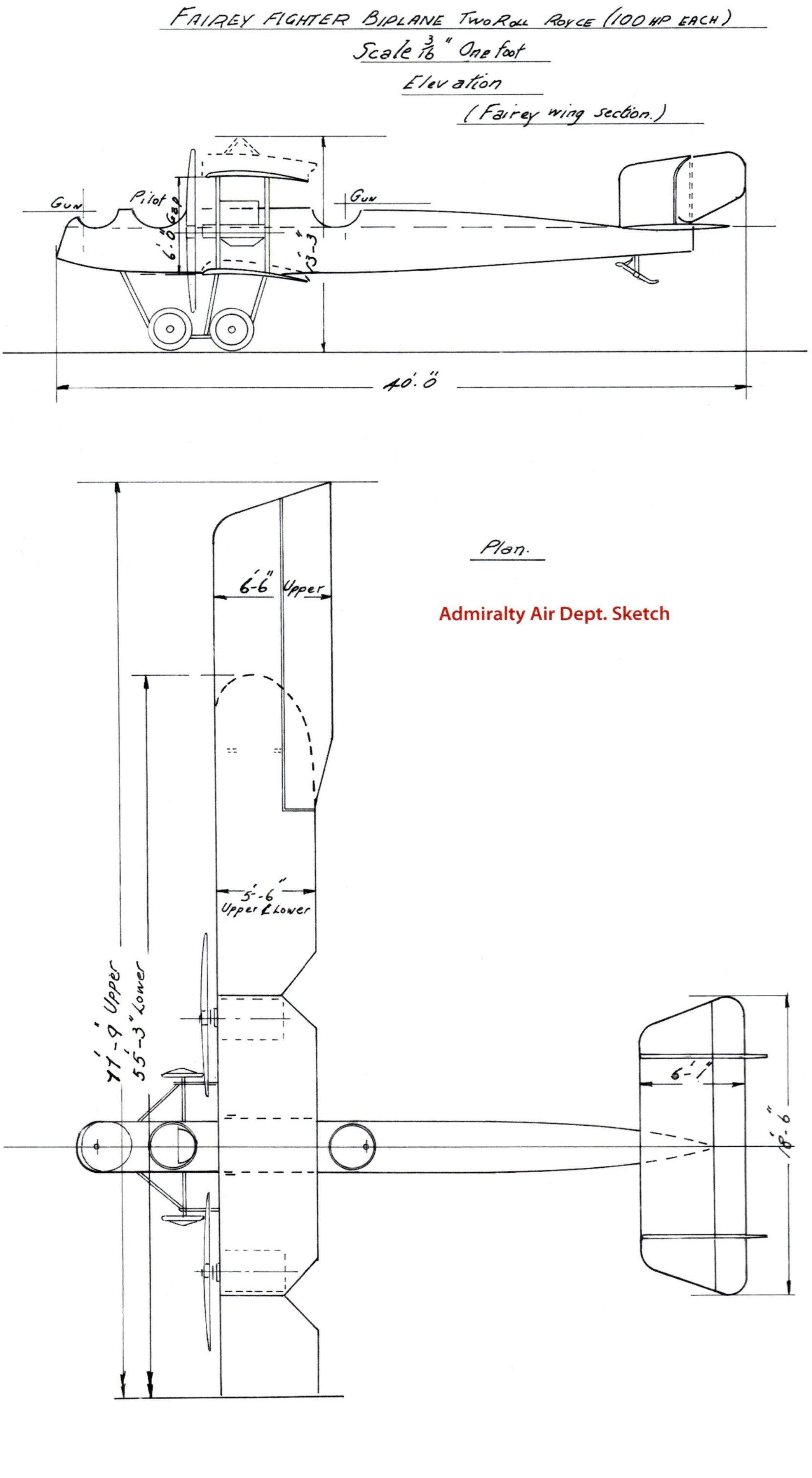

Admiralty Air Dept. Sketch

Above: The F.2 seen soon after assembly in the Harrington shop. The original five wooden sheds on the site are visible in the background. Note the tailplane support strut and lack of serial.

the two gunners, this will be permitted in lieu. The machine was to be capable of *cruising comfortably* at a reduced speed of 75 mph. A 250-hp Rolls-Royce engine would be supplied by the Admiralty for this machine.[1]

It appears from the replies from the aircraft manufacturing companies that around 4 December, twin-engine machines were included in the competition. Faireys received a letter that noted that the firm *may be informed that the designs to be submitted are not limited to single engine machines. If the design they wish to submit involves twin engines it will only be possible to supply either 140 or 160 H.P. Sunbeam engines, but at a later date it will probably be possible to supply 180 H.P. Rolls Royce Engines in lieu. These latter engines, however, will not be available till well after the middle of March 1916.*[2]

Mann, Egerton & Co Ltd of Norwich; Robey & Co Ltd of Lincoln; The British & Colonial Aeroplane Co Ltd of Bristol; The Blackburn Aeroplane & Motor Co Ltd of Leeds; A.V. Roe & Co Ltd of Manchester; Hewlett & Blondeau Ltd of Bedfordshire, and the Fairey Aviation Co Ltd all replied to the Admiralty's approach. Bristol did mention that their workload may not allow them to proceed with a design. Blackburn had their general arrangement drawings ready by 17 December 1915. Avro were to have two 160-hp and one 249-hp Sunbeam engines forwarded as soon as possible in February 1916.[3]

Fairey and Avro each received an order for two Heavy Bomb Droppers. Fairey's were given the serial Nos. 3702–3703. They were to be powered by two Brotherhood 200-hp engines. Brotherhood Ltd had made engines for the first Whitehead torpedoes. Judging from the surviving correspondence, the availability of these 200-hp engines would have been seized upon despite their manufacturer not being known as a supplier of aeroengines.

The Sections of the Air Department of the Admiralty presented weekly reports and some of these have survived and give some idea of the development of the Heavy Bomb Dropper.

The bomber is first reported for the W/E 10 February 1916, when Sqn Cdr Seddon visited Fairey at Hayes regarding the "Mock-up" of Fairey bomb dropper.

Controls and sighter arrangements found to be satisfactory and well thought out.

Anticipated performance adjudged to be far below present standards, and mater is being brought before construction sub-committee.

Despite this Fairey's tender for two Heavy Bomb Dropper Aeroplanes was received in the W/E 4 March, and the price was under consideration. The following week it was reported that a visit had been made to the Fairey Aviation Co in connection with the A.D. Bomb Dropper. The tender was considered excessive and the firm was asked to reconsider. For the W/E 31 March 1916, Y Section put forward a recommendation to cancel the order for the Heavy Bomb Dropper *it having been decided that these 2 machines are not up to requirements.*

Sturtivant and Page record that the two bombers were cancelled in March 1916, and again in November. For the W/E 7 April1916, it was proposed to *Order 2 Davis gun carrying machines of own design (from Robey & Co.). These will take place of 2 Fairey heavy bomb droppers which it is proposed to cancel. They will complete* (order) *CP 04921/15.* These were the Robey-Peters Gun-Carrier tractors ordered 31 May 1916, from Robey & Co Ltd. Allotted serial Nos. 9498 – 9499, only the first was to be completed.[4]

During the W/E 4 May, two visits were paid to Fairey Aviation Co in connection with the A.D. Bomb Dropper. These would most probably have been in connection with finalizing the project. It is assumed that work on the bomb dropper would have been carried over into the fighter.

According to Taylor, Nos. 3702 and 3705 did not get beyond wing-detailing stage when a requirement for folding wings was added to the specifications. He contends that these two aircraft were then not proceeded with. One was to be of pusher configuration and one a tractor. Sturtivant & Page show Nos. 3702 and 3703 as tractors. It appears that the Heavy Bomb Dropper (F.1?) never left the drawing board.[5]

The first mention of a fighter version of the aeroplane is in the report for the W/E 7 April 1916, when Lt Barlow of "D" Section, visited Faireys on inspection duty in connection with the "Fighter No. 3704." They were ordered on contract

Above & Below: The F.2 seen soon after assembly in the Harrington shop. The original five wooden sheds on the site are visible in the background. Note the tailplane support strut and lack of serial.

No. C.P.110731/16 as A.D. Pusher Biplanes with two 190-hp twelve-cylinder vee liquid-cooled Rolls-Royce Falcon engines according to Sturtivant & Page.

Work apparently commenced about the time the Heavy Bomb Dropper was abandoned. Lt Barlow visited Faireys with reference to the "Fighter (3704/05)" during the W/E 14 April. Barlow made another visit to Faireys with Lt Landells in the W/E 5 May. Landells appears to have taken over from Barlow as he continued to visit Faireys with reference to the Fighter No. 3704 throughout May.[6]

For the W/E 16 June 1916, Y Section: -*Put forward to inform Firm that with reference the two Twin Engined Gun carrying aeroplanes Nos. 3704/05, these should be built to the specifications and that a financial limit has been fixed.*[7]

The machine that emerged to become known as the Fairey F.2 was a tractor. The requirement for folding wings may have meant that it was not possible to achieve this in a pusher layout. The F.2 was a large three-bay biplane with the upper wing of greater span than the lower. The wing extensions were braced by triangular king-posts on the upper wing. The machine had a substantial four-wheel undercarriage to prevent overturning on landing, the forward wheels were of slightly smaller diameter than the rear wheels. When at rest it was supported by the rear wheels and a tail skid. The wings folded outboard of the engines that were mounted well up between the wings being supported by the interplane and separate support struts. The engines were handed in order to eliminate torque. The wide span tailplane carried twin fins and rudders. A small triangular kingpost was mounted halfway between the elevators. Armament comprised Lewis guns on Scarff ring mountings. Strangely Taylor states that bombs could be carried on external racks. Although the machine had the capability to carry bombs there is no indication that it was ever considered as anything but a fighter.

No. 3704 was delivered by road to the Design Flight, Eastchurch on 28 February 1917. Erected by the makers it was flown on 15 March. It then went to the Isle of Grain. Its first official flight by Sidney Pickles was on 17 May.[8] Too slow to compete with the single-seat fighters then redefining the air war, the F.2 was already obsolete.[9] On 12 May 1917, the had been spelt out as a *recommendation has been made to stop all work on the second machine in each case of the two experimental twin-engined tractor fighting machines ordered by the Admiralty last year from Messrs. A. V. Roe and Messrs. Fairey respectively. These machines are now of an obsolete type.*[10]

At the end of June 1916, Avro had been invited *to tender for*

Above: The Admiralty required folding wings for its landplanes as well as its seaplanes in order to fit them into the hangars of the time.

Above: The instruments on the inside of the engine nacelles are evident in this view. Note how the rear lower wing spar was left exposed where it carried across under the fuselage.

2 A.D. Fighting Aeroplanes Nos. 3694-3695 to subject to certain modifications.[11] This machine to emerge as the Avro 529. No role could be found for these large, heavy and relatively slow multi-seat fighters. With their endurance they may have been developed into an anti-Zeppelin fighter but were not further developed and the concept abandoned.

The second F.2 was not completed, only the wings having been constructed when the contract was cancelled.[12] In his usual fashion, having no regard to the facts, C.G. Grey of *The Aeroplane* recorded that in *spite of its excellent performance and its superiority to any other types then available, it* (the F.2) *was lost amongst the shifting sands of our aerial policy, and was never reproduced.*[13] Later he berated the Air Board and Air Ministry as fools or knaves: - *Were they fools or knaves who turned down the twin-engined Dyott and the twin Fairey fighter, and the twin Avro "Pike," each in its day better than any contemporary twin-engined machines?*[14] C.G. Gray had a close relationship with Richard Fairey. *The Aeroplane* of 17 November 1920, gave the F.2 a full page with photographs under the title "A Matter of Priority." As the editor, C.G. Gray, pointed out that while the Germans were supposed to have produced the first multi-engined aeroplanes of modern type, the "Fairey Twin-Engines Folding-Wing Bombing Biplane" was designed in 1914 and built in the Summer of 1916. Specifications given are the same as used

Fairey F.2 Specifications			
Source	**1**	**2**	**3**
Dimensions			
Span upper	77 ft	77 ft 9 in	76 ft
Span lower		55 ft 3 in	
Span folded			24 ft
Length	40 ft 6½ in	40 ft 0 in	41 ft 4 in
Length folded			42 ft
Height	13 ft 5 5/8 in	13 ft 3 in1	10 ft 7½ in
Chord	5 ft 6 in	5 ft 6 in	
Gap	6 ft		
Span tailplane	17 ft 6 in	18 ft 6 in	
Areas in ft^2			
Wings	718.4		684
Ailerons (each)	47.8		
Tailplane	64.9		
Elevators	39		
Fins (each)	10.9		
Rudders (each)	10		
Weights in lbs			
Loaded	4,880 /4680		5,628
Pilot & 2 crew			540
Fuel (84 gall)			609
Oil (6 gall)			60
2 x Guns & mounts. 7 x Double trays			145
Total useful load			1,354
Light (with water)			4,274
Speed in kts			
at 3,000 ft			78
at 5,000 ft			79
at 7,000 ft			73
at 10,000 ft			72
Max (GL)	81 kts		
Landing speed	38 mph		
Climb to			
3,000 ft			4 min
5,000 ft	6 min		7 min 10
10,000 ft			20 min
13,000 ft			37 min 15
Ceiling in ft			14,750
Endurance	3½ hrs		Approx. 2½ hrs
Engine			2 x 190-hp Rolls-Royce
Notes: 1. Measured to top king post.			

Above: The serial 3704 and the c/n "F".2 can be discerned on the original print. The photograph was taken when the F.2 was at Grain in 1917. The tailplane support strut has been omitted. The upper footstep has been omitted and a hanging step introduced.

Above: The F.2 at the Isle of Grain on 21 May 1917 with two-bladed airscrews and uncowled engines.

in *Jane's 1919,* Grey also being the editor of that preeminent annual.[15]

End Notes

1. M.F.S. (M.F. Sueter) Director of Air Services to Short Brothers, December 1915. TNA AIR1/664/17/122/705.
2. W.L.E. for Director Air Services to Lt Cox, RNVR, at Fairey Aviation Co. 2.12.1915. TNA AIR1/664/17/122/705. A Smith in *The Man Who Built The Swordfish – The Life of Sir Richard Fairey* (I.B. Tarus, UK, 2018) states that the contract was for "Twin Tractor Armoured Fighting Aeroplanes."
3. W.B. for Director of Services to Lt Bell, RNVR, at A.V. Roe. 14.02.1915. TNA AIR1/664/17/122/705.
4. Ordered 31 May 1916. Sturtivant, R & Page, G. *Royal Navy Aircraft Serials and Units 1911-1919,* Air Britain Publication, UK, 1992. P.163.
5. Smith, A. *The Man Who Built The Swordfish – The Life of Sir Richard Fairey,* I.B. Tarus, UK, 2018.
6. *Jane's All the World's Aircraft 1919* states that the machine was designed in 1914 which is incorrect. It also states that the F.2 was the first twin engine machine with folding wings. C.G. Grey was the editor of *Jane's* at this time.
7. This section has been written in the main from A.D. Weekly Sections Reports February to June 1916. TNA Air1/337/15/226/246.
8. Sources give different dates for the first flight. This is thought to be the correct sequence. C.G. Gray's assertion that the machine was completed and flew in 1915 is false.

Sources for Fairey F.2 Specifications on facing page:

1. *Jane's All the World's Aircraft 1919.* Same data in J.M. Bruce, *British Aeroplanes 1914–1918.*
2. AD drawing of "Fairey Fighter Biplane Two Rolls Royce (190 HP each)".
3. Controller, Technical Dept Report of 23.05.1917. Copy in RAF Museum J.M. Bruce Collection Box 25.

Above: The F.2 at Grain 21 May 1917. Note the rudder control positioned in the centre of the fixed horizontal tailplane. The cross bar is parallel with the rudders.

Above & Below: The F.2 appears to have its engines uncowled in these views, probably on the Isle of Grain. Note the triangle shaped kingpost on the tailplane. The tailplane bracing was elaborate. There were no control horns on the rudders and they were operated from the extensions of the steel pipe that the rudder rotated about that protruded below the horizontal tailplane.

Above: This view of the crashed F.2 was taken at the Isle of Grain. The F.2 was there from March to June 1917. Chassis repairs are recorded for August 1917; however, this photograph shows that more than the chassis required repair. The aeroplane was finally written off in the W/E 23 February 1918. What, if anything, it did in the interim, is unknown.

"Modern British Aeroplanes X – The Fairey Aviation Co. Ltd." *Aeronautical Engineering,* Supplement to *The Aeroplane* of 6 August 1929. P.509.

9. A Smith states that although it was an impressive effort in terms of manufacture and design, the F.2 was a year overdue on its delivery date and obsolete. He seems to have accepted the story that the F.2 was designed in 1914. *Op Cit.* P.82.
10. Technical Dept report of 12.05.1917. Signed by Brig-General D Pitcher. Copy in RAF Museum, J.M. Bruce Collection Box 78.
11. TNA Air1/337/15/226/246.
12. Fairey adopted the construction number (c/n) of the first aircraft of a new type or major modification as the designation for that aircraft. Thus, the Fairey Campania became the Type F.16, F.17 and F.22. The actual designation of the first Fairey aircraft to be built as the Type F.2 was published in *Jane's All the World's Aircraft 1919.* Given the relationship between *Jane's* and the aircraft manufacturers, this is assumed to be correct. The entry is illustrated by photographs of No. 3704. This would mean that No. 3702 was given the c/n F.1, No. 3703 was not given a c/n and therefore No. 3704 was F.2. As shown in the Appendix in Taylor, (Taylor, HA. *Fairey Aircraft since 1915,* Putnam, UK, 1974.) aircraft contracted but not built were given a c/n. Taylor has confused the Heavy Bomb Dropper with the Fighter and has assigned the c/n F.1 to No. 3702 Fighter (not built), and F.2 to No. 3704. Sturtivant and Page (*Royal Navy Aircraft Serials and Units 1911-1919,* Air Britain Publication, UK, 1992.) assign F.2 to No. 3702 and proceed from there in numerical sequence. Thus No. 3704 is c/n F.4. However, they then assign F.4 to the first Short 827 Type, the same as does Taylor - (See Sturtivant & Page P.136). The question then is whether No. 3704 was in fact the Fairey F.3 and the designation given by *Jane's* as the F.2 has become accepted over time. The simple answer may be that the assignment of c/n may have been adopted with the order for the Short floatplanes at c/n 4 for No. 3550 and retro-allocated to the earlier aircraft.
13. "Modern British Aeroplanes X – The Fairey Aviation Co. Ltd." *Aeronautical Engineering,* Supplement to *The Aeroplane* of 6 August 1929. P.509.
14. "On the Flying Services, the Aircraft Industry and Graft." Editorial, /4680 *The Aeroplane,* 29 October 1919, P.1500.
15. Fairey was one of the few British manufacturers that had all its aircraft built during the war described in detail in *Janes's 1919.* For the relationship between Grey and Fairey see Smith, A, *Op Cit.*

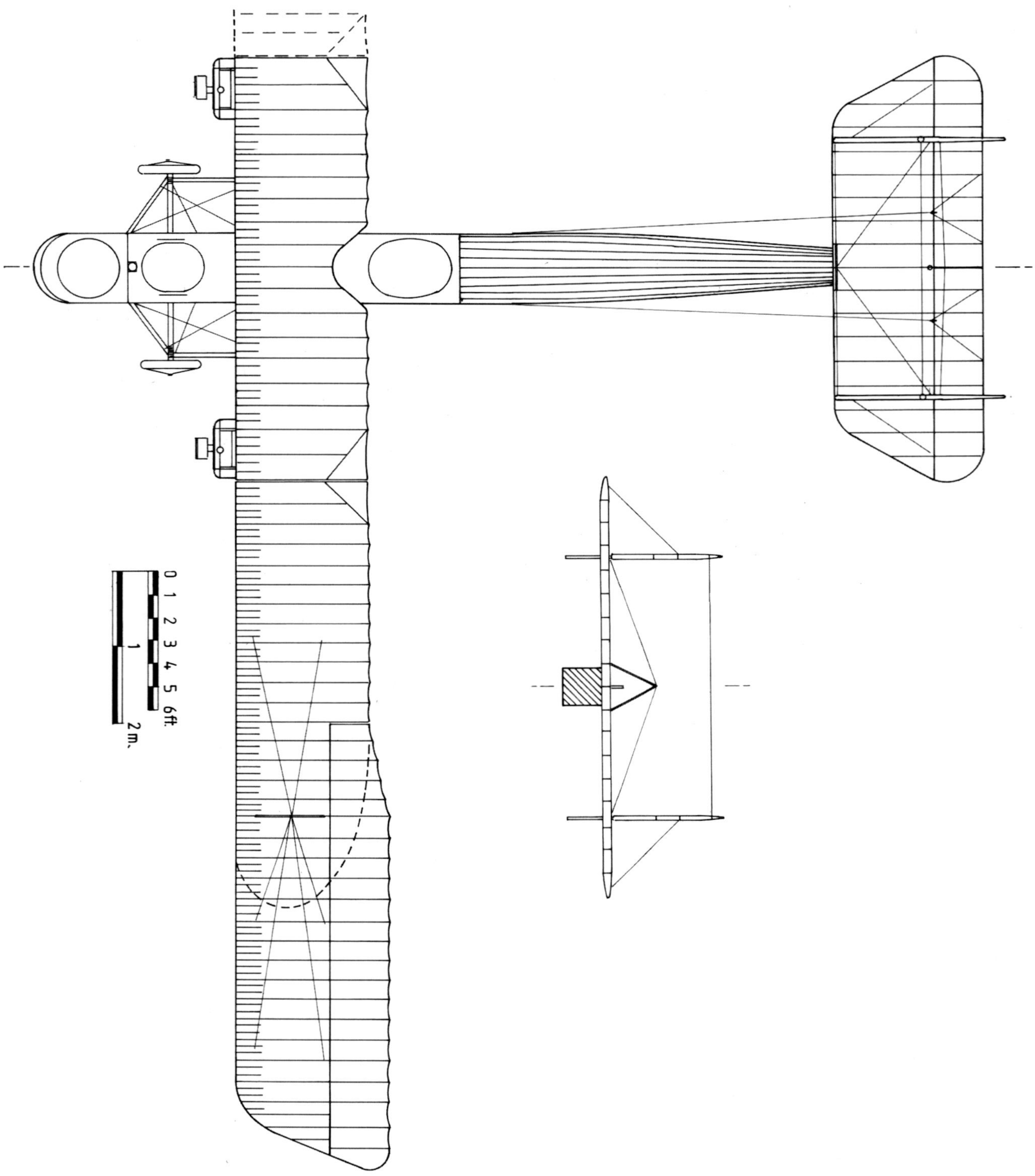
0 1 2 3 4 5 6ft.
1 2m.

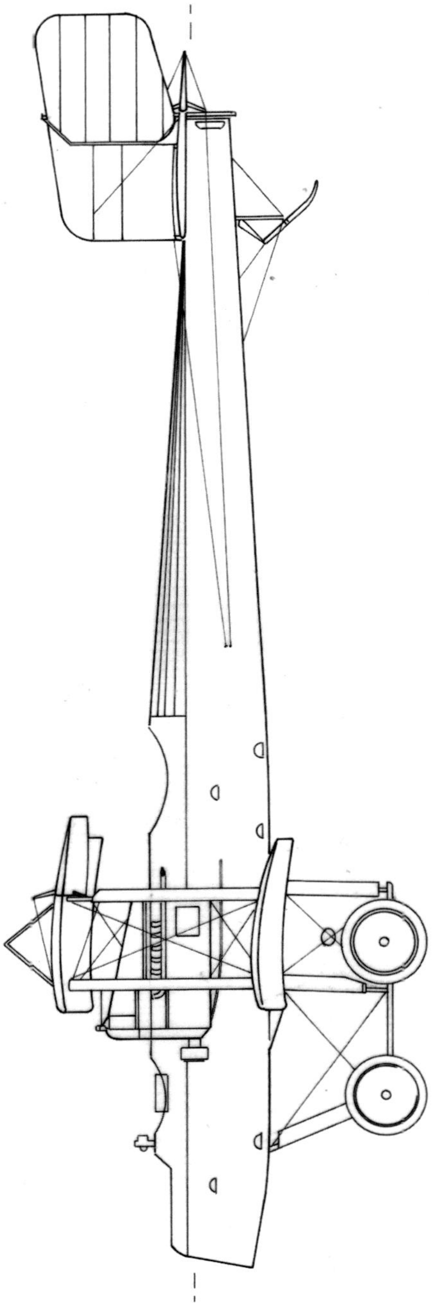

Tail plane omitted for clarity.

Chapter 3: The Fairey Campania

Above & Facing Page, Above: N1000 as first rolled out at Hayes. Note the exhaust pipes for the 250-hp Rolls-Royce Mk. IV (later Eagle IV) pass through the centre-section of the upper wing. The chin radiators are well shown.

The Fairey Campania is one of the least known floatplanes of WWI despite being a handsome and practical machine for its time. The Campania came about due to the Admiralty's requirement for an aircraft to operate from HMS *Campania.*

After the outbreak of the war the irrepressible Capt Murray Sueter of the Air Department of the Admiralty suggested that an old passenger liner be converted to a seaplane carrier that could keep up with the Grand Fleet as the Royal Navy's seaplane carrier, HMS *Ark Royal,* could only do 10.6 knots. The old Cunard liner *Campania* had been constructed in 1893 for the North Atlantic service, and had a speed of 22 knots and, as one of the few ships capable of meeting the requirements for a seaplane carrier, she was purchased by the Admiralty in October 1914, for this purpose. The ship was modified by Cammell, Laird between November 1914 and April 1915. The modifications included a flying-off deck, 120 ft in length, above the forecastle, and provision for ten seaplanes. As no method had been devised for allowing floatplanes to take-off from the deck, early flights were conducted by lowering the seaplanes to the water, but this proved to be costly in smashed floats or aircraft, and a wheeled trolley was developed for use with the flying-off deck, these being discarded as soon as the seaplane became airborne. Later a method was used that allowed the trolleys to be recovered for further use. The seaplanes still had to alight on the water and be picked up by the ship's crane.

A Sopwith Schneider floatplane piloted by Lt W.L. Welsh, made the first successful take-off using a trolley from *Campania's* deck on 6 August 1915. It required 113 feet for the take-off run, and it was apparent that no larger seaplanes would be able to utilise the deck. The *Campania* returned to have the deck extended to 200 feet. Provision was also made for a kite balloon to be carried aft. The first flight by a two-seat seaplane from her deck was on 3 June 1916.

In late 1916 the Fairey Aviation Co was asked to design a two-seat floatplane capable of operating from ships such as the *Campania.* The dimensions of the ship's hold were the critical criteria for the size of the seaplane. Folding wings were mandatory. By October work on stress checking the design was in hand.[1]

F.16/N1000

Fairey works number F.16 was applied to the prototype that received the serial N1000[2]. Power was provided by a 250-hp Rolls-Royce Mk. IV (later Eagle IV) twelve-cylinder vee water-cooled engine. The radiator was mounted in two blocks, one on each side of the engine. The exhaust pipes were raked backwards to pass through the upper wing centre-section. Ailerons were mounted on the upper wing only. The wings

Above: N1000 thought to be still at Hayes but with a four bladed airscrew.

Above & Below: N1000 thought to be still at Hayes but with a four bladed airscrew.

were designed to fold, and when folded the base of the outer rear interplane strut engaged a clip on a stay that projected out from the side of the fuselage. The upper wing was of longer span than the lower and the extensions were braced by triangular king posts situated above the outer interplane struts. The floats were of contemporary design, similar to those of the Short floatplanes. The deep tail float was attached directly to the fuselage lower longerons. The wingtip floats were flush against the lower wing surface and sloped outwards on their bottom surfaces to direct water away from the aircraft.

Armament was a single Lewis Gun on a Scarff ring for the observer. Bombs could be carried on an external tubular rack under the fuselage.[3]

Because of Richard Fairey's work for Short Brothers it was natural that the Campania showed influence of the Short floatplanes, but Fairey's chief designer, F Duncanson, incorporated his and Fairey's own ideas, such that the Fairey seaplanes soon overtook Shorts and post-war were the prime builders of floatplanes while Shorts developed flying boats.

Sidney Pickles conducted the first flight of N1000 on 16 February 1917, at Hamble. A Report on the trials of N1000 was published in the 20th Weekly Report of the Controller, Technical Department for the W/E 25 April 1917.

N.1000 Fairey "Campania" type:

Speed, low down,	*67.7 knots,*
Speed at 3,000 ft.	*67.5 knots,*

Above & Below: N1000 at the Isle of Grain. Note the four-bladed airscrew.

Above: N1000 at the Isle of Grain. Note the four-bladed airscrew.

> *Flying to 6,000 ft. in 24.5 minutes.*
> *Total Military Load, 1,448 lbs. including fuel for 8 hours at full speed or 10 hours at cruising speed.*

> *This machine is intended for Reconnaissance work only and carries no weight of bombs.*[4]

A chart of "Seaplanes in use in RNAS (Not including experimental machines)" described the Fairey Campania 250-hp Rolls-Royce as a *2-float seaplane designed for use in seaplane carriers. Carries 8 hours fuel, and is to be used for fleet reconnaissance. First machine just completed trials, and will not be available in quantity for some months.*[5] The chart is not dated however, it confirms that the machine was to be assigned to fleet reconnaissance and not anti-submarine patrolling.

The "Fairey Experimental Type II A Engine 190 h.p. Rolls-Royce" was reported under test in May and while it appeared to be promising, its speed was likely to be less than that of the A.D. boat.[6] The same report noted that the Campania "Type IIb 275 hp RR" was being erected preparatory to test.[7] It was proposed to compare the two float seaplanes with the A.D. boat as *they are both carrying the same load.*[8] It thus appears that the A.D. flying boat was being presented as a fleet reconnaissance machine. It was not until July that N1000 underwent its official trials at the Isle of Grain.

F.17/N1001

The second Campania was powered by a 275-hp Rolls-Royce Mk. I engine (Eagle V). While the principal difference was the type of engine installed, there were considerable differences between the two prototypes. No photograph of N1001 is known to have survived. Campanias N1002 – N1004 had the same engine installation and photographs of N1004 have survived and it is assumed that N1001 would have been similar. The engine was mounted further forward and the exhaust pipes no longer passed through the wing structure. A more efficient aerofoil section was used. The centre-section was reduced in chord to improve wing folding, and had a large cut-out in the trailing edge to improve the pilot's upward vision. The float undercarriage remained unchanged but the wing-tip floats were now braced by short struts. Most Eagle powered Campanias, including N1004, had a frontal radiator, however the Technical Notes for the Fairey Campania refers to side radiators with the Rolls-Royce Eagle engine.[9]

As noted above, the "Fairey Campania type II B 275 h.p. Rolls-Royce" N1001, was reported as being erected in the Minutes of a meeting in May; however, a report of 26 April 1917, noted that the trials of N1001 were nearly completed. The machine was described as having a good performance and was in balance. Although a full report had not been received the top speed was stated to be 74.8 knots at 1900 revs at sea level. The airscrew was considered not suitable for the trial and trouble had been experienced with the oil overflowing due to overheating.[10] The US Navy noted that same month that the Campania type with "250 RR Engine" was being developed with increased speed and fuel endurance, having passed the experimental stage.[11]

Both N100 and N1001 were to be later used at the Scapa Flow RNAS station.

The initial order was for ten machines (N1000 – N1009) included the two prototypes. Production of the seaplane as the Type F.17 commenced from N1002. Production aircraft followed N1001 rather than N1000. The wing folding mechanism was modified with a fitting engaging the base of the

Above, Below, & Following Two Pages: Although no photographs of N1001 appear to have survived, it is thought that N1001 was similar to N1004. The 275-hp Eagle V was situated further forward with exhaust stacks that avoided the centre-section. The production machines followed F.16/N1001 and it is most probably had a nose radiator such as that shown here on N1004, the earliest one to have survived in photographs. The fin and rudder were larger than N1000. Photographed at the Isle of Grain on 29 September 1917. Note the four-bladed airscrew, the enlarged ailerons and the wing-tip floats are now suspended below the wings on three struts.

inner rear strut on each side. Although Fairey drawings showed ailerons projecting beyond the trailing edge of the wing, some of the machines in the N1000 - N1009 batch had flush ailerons the same as N1000. All later Campanias had a larger fin than N1000 and rectangular kingposts.

N1002 was taken up by Sidney Pickles at Hamble on 3 June 1917. It was at Grain and being erected later that month for service trials, after which it was to be flown away to *Campania* at Scapa Flow by Lt Maurice E.A. Wright.[12] At the Air Department Progress Committee meeting in May, Capt A.V. Vyvyan[13] wanted Wight No. 9849 to be flown to *Campania* after it finished its trials at Grain. This would appear to have been required in order to test the two different seaplanes operating with the *Campania*. Later that month it was reported that N1001 had been successfully flown to the *Campania,* the machine *appeared to be in all respects satisfactory except there may be a slight trouble with the floats which were weak.* Arrangements were being made for the Fairey Company to send a spare pair to *Campania.*

In June 1917, the question of future orders for the Fairey Company arose and it was suggested that more Campanias should be ordered; *the only difficulty was the supply of Rolls-Royce engines.* It was suggested that engines be obtained from wrecked or deleted aircraft, however it was pointed out that these engines had to go back to the maker before reissue and very little time would be gained from the proposed procedure. As an alternative it was put forward that the R.A.F.3a engine be used in the Wight Converted Seaplanes on order, and the Rolls-Royce engines previously allocated to these aircraft, would then

Above & Facing Page: The cockpit of Fairey F.20 Campania, 275-hp Eagle powered N1004, 29 September 1917. The Patten 259 compass was inside the windshield. Note the Fairey plate on the port side of the dashboard; and the altimeter appears just above the centre of the wheel control column, with the inclinometer situated directly underneath the altimeter. To the right are the Oil pressure gauge, Hollocombe-Clift ASI (in knots), and the Rev counter.

be made available for the Campanias. This proposal depended on whether the Wight machine was a success with the RAF engine. *This experiment was being hastened as much as possible and arrangements had already been made for an R.A.F. engine to be sent down.*

The Fairey design was thought to be the better machine, but the Wight had not performed any proper trials. It had been accepted as a successful type on the *strength of an enthusiastic report from Calshot, but there now appeared to be a slight doubt as to the strength of certain parts and* (Crd Longmore) *was making arrangements for one of these machines to be sent to Grain Island for a through testing by the Test Flight there as soon as possible.*[14] The Wight never had a RAF 3a installed, and a 260-hp Sunbeam was proposed to be installed instead.

The next Meeting of the Committee was informed of the results of the trail of a Campania. This machine had a speed of about 8 knots higher than the "Short." The first one had just been delivered to the *Campania* by air. The only trouble was the very large diameter airscrew that only had a very small clearance. It was pointed out that the Rolls-Royce engine that would be fitted towards the end of the contract for 70 machines now on order would be of considerably higher horse power than the present 275-hp type, *and if trouble were* (sic) *already experienced due to the large propeller required for the 275 H.P. engine* (Lt Andrews) *was afraid that even greater trouble would be experienced when the higher horse power engines began to arrive.* It was pointed out that it was hoped to get out a four-bladed airscrew that would relieve the present trouble. It was noted that as far as possible the higher-powered engines should be allocated to the Large America flying boats.[15] Rolls-Royce engines were also required by the Army and were always in short supply throughout the war.

The Committee's regard for the Fairey seaplane can be seen in the discussions regarding the Baby seaplanes. These single-

seat fighters were no longer up to their work and were generally out-classed by the latest German machines. It was proposed that the order for these be replaced with one for Sopwith Ships Camels. Cdr Longmore proposed that instead of Fairey being given an order for Camels, the firm should be given an order for more Campanias, the engines for these coming from those allocated to the Wights. The Wights would utilize the 260-hp Sunbeam engine. If these proposals were approved then a further 40 Campania seaplanes were to be ordered.[16]

In July the Fairey Campania was *the best type we have at present but there are no more Rolls-Royce engines available for any further orders* and it was necessary to try the machine with the 260-hp Sunbeam *otherwise we should be reduced to ordering still more of the 184 Type Short*. The trying out of the Sunbeam engine was a priority, *so that when it becomes necessary in a few weeks to place further orders with various subcontractors now building Shorts, orders can be placed for the Fairey with the 260 Sunbeam rather than for more Shorts.* N1006 was selected as being an advanced machine for the modification as this machine was just far enough advanced to enable the modifications to be effected with the least delay. If at all possible a new engine was to be sent to Faireys, one with all the latest modifications, such as new pistons, etc.[17] Fairey were to receive five 250 Rolls-Royce, ten Series 8 Rolls-Royce and 25 Sunbeam Maori engines for their contract for 40 machines (N2360 - N2399).[18] The last 25 of this batch were to receive the Sunbeam engines.[19] As at 7 January 1918, Fairey had delivered 13 Campanias from this contract.

At a Meeting of the Weekly Progress Committee in August, Cdr Longmore was able to state that the *Campania* was pleased with the "Rolls Fairey Campania," but had enquired why the fuel had been cut down to 4½ hours. This tankage was made standard for the machine; however, it was suggested that spare tanks could be supplied to fit in the space originally provided for the 35-gallon tank.[20] Stations were allowed to hold larger tanks for use when operational requirements required greater endurance.

Barclay, Curle & Co, Ltd, a ship-building firm of Whiteinch, Glasgow, received an order to build 50 Campania floatplanes (N1840 – N1889). In August 1917, the firm still required a considerable number of Campania drawings, but had enough to start work, sub-contracts having been placed.[21] In late August it was noted that there were only sufficient Rolls-Royce engines for the first 25 Campania seaplanes ordered from

Above, Below, & Facing Page: N1004 was with the Type Test Flight at Grain and came to a sticky end at the Isle of Grain on 14 April 1918. As the floats are under the fuselage it appears that the machine managed to avoid hitting the embankment but stalled into the ditch. Details of this accident have not been discovered to date.

this company due to demands from the Army, and the 260-hp Sunbeam would have to be used for the remainder of the order. The experimental installation in N1006 was almost complete and the necessary drawings would be ready for issue as soon as required.[22] The first Barclay, Curle Campania was not delivered until 30 March 1918.

Later that month questions were asked about the fitting of tubular radiators to the Campania. The frontal radiator was stated as satisfactory for home service; however, it was proposed to send the type to the Mediterranean and the radiator required

Above & Below: Details of the Sunbeam Maori II engine installation in the Fairey Campania. The central exhaust was a characteristic of the Sunbeam Maori II engine.

Above: Campania Sunbeam Maori engine cowling. Note the central exhaust pipe.

Below: Fairey F.22, 250-hp Sunbeam Maori II powered N1006, on the beach at Calshot, December 1917. Note the petrol tin on the float. It appears that the machine is being refueled. N1006 was the first Campania seaplane to receive the 250-hp Sunbeam Maori II engine. Note the small radiator without shutters.

Above: N1006 at Calshot in December 1917. Note the Wight seaplane in the background. N1006 had brief stays at Calshot and Dover before reaching Portland by 21 December 1917. It was deleted at this station during the week ending 24 July 1918.

Above: Fairey Type F.22 Campania N1006 at Calshot 13 December 1917. Note the flush ailerons.

This Page: N1006 photographed at Grain. This seaplane served at Calshot and Portland in No. 415/416 Flight, May 1918.

Above, Below, & Facing Page: Fairey Type F.17 Campania N2363 with 275-hp Rolls-Royce Eagle VIII, at the Isle of Grain, 10 December 1917. The engine installation featured revised cooling panels and louvers Note the full bomb load, four-bladed airscrew. N2363 was delivered to Grain's Type Test Flight on 10 December 1917, and later served at Calshot and Portland.

would have to be larger. *The honeycomb radiators mounted in front could not be enlarged when required, as could tubular radiators, by the addition of extra sections as required, and it was for this reason that the latter type is being adopted.*[23] N1002 was at Grain undergoing tests and the cooling surface of the radiator was not enough, and it was to be increased by an extra row on each half radiator. It was suggested that a honeycomb radiator be designed in place of that submitted by the firm.[24] N1004 was fitted with a special honeycomb radiator by Fairey and was tested at Grain, this radiator became standard for all Rolls-Royce powered machines.

The installation of the Sunbeam Maori engine led to the Admiralty requiring a larger radiator. It appears that Fairey raised objections that enlarged honeycomb radiator would be heavier, but the reply was that the Campanias were to be sent to the Mediterranean and the larger radiator was therefore required. Shutters were fitted to the lower half of the enlarged radiator which became standard for Maori engine machines. However, no Campanias ever served in the Mediterranean.[25]

Cdr Longmore reported in October that the trials of the first Campania with the 260-hp Sunbeam engine, N1006, were not satisfactory, being inferior to the 184 Short with the same engine. A minute had been sent to the Controller of Aeronautical Supplies asking that orders for Fairey machines with the various sub-contractors should be altered to Shorts – *for the present – until the Fairey machine had been got right.* Barclay Curle's order for Faireys would not be affected as Rolls-Royce engines were available for the first 50 of this contract; and the last 25 would be fitted with Sunbeams. *Messrs. Fairey would be given every opportunity of making any necessary alterations in N.1006, as it is essential that they should get the machine right before they begin to fit Sunbeam engines in the last half of their contract.* It was stated that there would be difficulty in obtaining sufficient 275-hp Rolls-Royce engines for both Barclay Curle and Fairey. Some of the older 250-hp Rolls and repaired engines sent back into service would have to be utilised. As far as possible, the 275-hp engines would be sent to Barclay Curle, and the 250-hp to Faireys.[26]

Mr Fairey said that there was no doubt that the performance of the Sunbeam Campania could be improved by using a different wing than those fitted to the Rolls-Royce machines. The only trouble would be that parts would not be the same and there would be some slight changes in the production at the Fairey works. Cdr Longmore was prepared to accept these differences provided that the Sunbeam engine *gave a really good performance, in which case production orders would follow of that type in preference to the 184 Short.* The two-subcontractors would be given interim orders for further 30 Short machines in order to keep them in production during the two-months it would be required to get the improved Sunbeam Fairey floatplane out.[27] Orders for 50 Campanias each were given to Robey (Requisition No. Br 177 A.S.25835) and Brush (Requisition No. Br 177 A.S.25836) before placing a formal contract on 7 September.[28] The following month, Mr Fairey asked which type of wing was required for the Sunbeam Campania, the ordinary type or the high lift wing *as at present fitted to N.1006 at Grain.*[29] A note to the file stated that the ordinary Campania wings were to be used.[30] J.M. Bruce has recorded that the second Campania, the F.17 version, used a more efficient aerofoil section than that used by the first Campania, the F.16.[31] This Meeting also considered that fitting two 230-lb bombs instead of four 100-lb bombs was "very desirable" and the 60 machines with the Rolls-Royce Eagle VIII should be able to carry this load.

The Weekly Report of the Air Department officer at the Fairey Company for 21 November 1917, noted that the drawings were completed except for the cowling, radiator and radiator mounting. Fairey was preparing drawings for the installation of the Rolls-Royce Eagle VIII engine. It was understood that the only other contractor building the Campania was Barclay Curle, and shortly before their first machine was completed, they were informed in February 1918, that they would receive all Series VIII drawings as their floatplane would have the Eagle VIII engine. The drawings

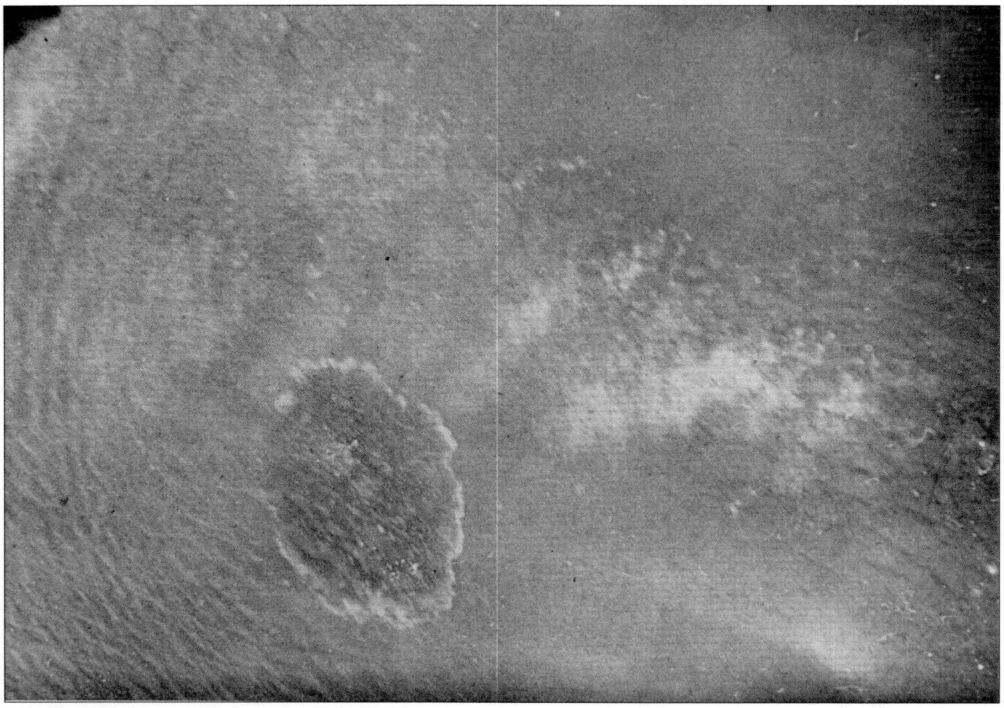

Left: Lt. Brewer, with Lt. Gauntlett as observer, dropped bombs on an oil patch from Campania N2363 on 12 June 1918. This view shows the bombs exploding.

Facing Page, Above: N2368 was operated from HMS *Pegasus*. The fuselage codes probably identified the ship but this is only supposition and more research is required on this subject.

Facing Page, Below: A dazzle painted HMS *Pegasus* under way with a Campania floatplane on the rear. Note the two cranes for lifting the seaplanes aboard.

Above: Cockpit of Campania Type F.17 N2363. 29 September 1917.

L.B.9
SEA PLANE

Above & Below: Two views of dazzle painted HMS *Pegasus*. The Fairey Campania at the stern appears to have the same fuselage markings in each photograph.

were being revised and corrected. These corrections were being sent through from Faireys and the tracings corrected at the Air Board.[32] Fairey received a second contract and deliveries began in November 1917. In December it was noted that the Fairey's contract for Campanias would be completed in three months. Barclay Curle only completed 12 of the F.22 Type Campanias with the Sunbeam Maori engines. They were delivered from March 1918 for allotment to the Grand Fleet.

The Fairey Campania Type F.22

The Campania had a slender fuselage of conventional wood and wire construction with fabric covering. Metal panels enclosed the engine. Rolls-Royce powered examples had twin exhaust stacks that sloped slightly backwards. Sunbeam engine powered examples had a single central exhaust.

The pilot sat behind the centre section that was cut back to give him upward visibility and allow for the wings to fold.

Behind him was the gunner's cockpit with a single Lewis gun on a Scarff ring. Although it was originally to be a fleet reconnaissance seaplane and not to carry bombs, they were fitted with a central bomb carrier below the fuselage similar to that used on the Short 184. It had been decided that the 60 Fairey Campanias with the Rolls-Royce Eagle VIII should be fitted to carry two 230-lb bombs.[33] The 260-hp Sunbeam Short 184 could carry one 230-lb or two 100-lb bombs.[34]

The main floats were twin pontoon-type floats and the upper surface of the main floats sloped 2° downwards towards the rear. They were braced by three pairs of bracing wires. Shock absorbers composed of one-layer of $^{7}/_{16}$ inch diameter elastic wrapped seven times were provided in two places at each end of the front float crossbar. Similar absorbers that comprised two layers of $^{3}/_{8}$ inch diameter elastic cord were provided for the rear crossbar. Each wing float was attached by means of three vertical steel tubes. These could be slightly adjusted by means of the front cross bracing wires and longitudinally by the side bracing wires. A large tail float was attached directly to the rear fuselage and incorporated the water rudder that was attached to and controlled by the rudder. The wing tip floats were attached by means of three vertical oval steel tubes.

There were three pressure fuel tanks. A 11 gallon (top) and a 78 (bottom) tank between the pilot and engine. Behind the engine was another 8 gallon tank. A compressed air starter was installed and ease of starting was stated to be good, the machine could be got away in 10 minutes.

The two-bay wings had extensions to the upper wing braced by kingposts. The main planes were hinged at the root of the rear spar so as to fold back along the sides of the fuselage, a jury strut being inserted near the front inner strut to keep the roots of the main planes in position when they were folded. When folded the main planes were secured to the sides of the fuselage by a sheet steel catch, fitted on the rear outer interplane strut. This consisted of a pin that engaged a slot in a small fitting attached to the bottom longerons near the gunner's cockpit. The catch on both sides could be released by means of a cable pull fitted on both sides of the pilot's cockpit.

Kingposts were fitted above the outer interplane struts on the top surface of the upper main planes. Each extension of the upper wing was supported by four bracing wires from the top of the top of the kingpost to the front and rear spars respectively. The kingposts were braced by two similar wires that ran to lugs on the upper attachments of the front and rear intermediate interplane struts.

The production Campania had the incidence of the upper mainplane at 5° throughout. The starboard lower main plane was set at 5°, that of the port lower main plane was 5° at the root, 4¾° at the inner interplane struts and 4½° at the outer interplane struts. This was to counteract propeller torque. Likewise, the incidence of the tailplane varied with the engine installed.

The tailplane was braced on the underside on each side of the fuselage by two 1½ inch x 0.6 x 22G oval struts, flattened at their ends and ferruled at their ends "by the standard method". On the upper surface the tailplane was braced on each side by two bracing wires, one from the top of the rudder post to the rear spar, and the other extending from the fin to the front spar.

With the 275-hp Rolls-Royce the tailplane was set at 4° for the model with side radiators and 3° with front radiator, 2° with the Eagle VIII and ¾° with the Sunbeam Maori installed. The side radiators were not to become standard as the Campania was never sent to the Mediterranean which was the rational for the side radiators, yet they were included in the "Rigging Notes."

Home Waters

The *Campania,* when she could be spared from working with the Grand Fleet, provided all the aircraft and personnel for the Scapa station. When enemy submarines were expected to be passing through the Fair Island Channel, seaplane patrols were made some going as far north as the Shetlands. She would occasionally be sent to a small anchorage in the northern islands of Orkney to form a base closer to the patrol area.

HMS *Pegasus* (previously the steamer *Stockholm*) and HMS *Nairana* (previously the Australian mail-steamer *Nairana*) also operated the Campania, however these smaller ships had flight decks that would not allow the two-seat floatplanes to be flown off. They had to stop and hoist the machines over the side. When not at sea their aircraft operated from shore bases such as Dundee, Newhaven, Rosyth and Scapa Flow. under Admiralty policy these ships were usually allocated the Rolls-Royce-powered Campanias. Those with the Sunbeam Maori engine were allotted to land bases such as Calshot (No.240 Sqn and No. 210 Training Depot Station); Bembridge and Portland (No. 241 Sqn).[35]

Unlike the Short 184, the Campania seaplanes did not take part in any spectacular actions. When the Grand Fleet departed for what was to become known as the Battle of Jutland, *Campania* did not get the signal to deploy and did not sail until some two hours later. Because of her speed she could have caught up with the fleet, but was ordered to return to base because she was unescorted and German submarines were reported in the area. This action was condemned after the battle; however, *Campania* had missed its chance for action.

The Campania floatplanes task was the endless searching on patrol for enemy submarines, mines, and escorting shipping.

Occasionally the monotony of patrol was broken. On 10 March 1918, N1009 from *Campania* was piloted by Flt Sub-Lt Kerruish, and whilst on patrol shortly after 1600, sighted a wake about 8 miles ahead and turned to investigate. It soon became apparent that the wake was caused by the conning tower of a submarine just breaking the surface. The seaplane proceeded at about 80 knots to attack, but before the objective could be reached the submarine began to submerge, and disappeared at 1615. Twelve seconds later two 100-lb bombs were dropped from a height of 800 feet about 250–300 feet ahead of the swirl which was still discernible, but only one bomb exploded. The machine circled the vicinity at a height

Above: Fairey Campania meeting with a capital ship.

less than 500 ft for a period of 40 minutes, but no oil or wreckage was observed, although it is considered that the bomb exploded at or near the submarine. Very's lights were fired indicating the presence of the sub, but the destroyer apparently did not see them owing to the brilliancy of the sun. Directly after dropping the bombs communication was established with destroyer G.21 by Aldis lamp, informing her that a submarine had submerged in a position approx. 59V steering east. The seaplane altered course for base at 1655. At 1705 another destroyer approaching from southwards was informed of the position of submarine, the seaplane then returned to base, alighting at 1824.[36]

Portland based N2363, piloted by Lt F.D.J. Silwood with Lt E.A. Sawyer, dropped two 230-lb bombs on a suspected U-boat on 16 May 1918. By 1918 it was realised that the 100-lb bombs were not capable of doing much damage to a submarine unless a direct hit was achieved.

On 31 October 1918, the RAF had 26 Rolls-Royce Eagle and 16 Sunbeam Maori powered Campania floatplanes. Of these the Grand Fleet had 24 Eagle machines the rest serving at shore stations.

Russian Intervention:

The Allies intervened in Russia after the October Revolution. The peace treaty that the Bolshevik government signed with Germany allowed the Germans to move troops From the Eastern to the Western Front and freed some 1,500,000 Austro-Hungarian and 100,000 German prisoners. The Allies supported the White Russians in order to keep Russia in the war and to protect the material that they had sent to Russia from falling into German hands.

HMS *Nairana* and her aircraft were involved in the campaign in 1918 and 1919. Originally laid down in 1914 as a passenger ferry for an Australian company the ship was requisitioned by the Royal Navy and converted into an aircraft carrier. She had a flying-off deck for wheeled aircraft but could not operate floatplanes from the 95-foot deck. She saw little operational service as she was employed as a training vessel and in carrying aircraft to ships that had flying-off decks. When she was attached to the British forces involved in the North Russian Campaign, she and her aircraft were actively involved in what has been described as probably the first combined air, sea and military operation when the Allies attacked Modyugski Island. The force comprised French troops, some Poles who were recruited in Murmansk, Royal Marines, a detachment of US Marines, with the French cruiser *Amiral Aube,* HMS *Attentive,* a fleet of mine-sweeping trawlers and the aircraft carrier HMS *Nairana.*

Nairana's operated five Campanias (N1844, N1847, N1848, N1849 and N2370) and two Sopwith Baby floatplanes (N1440 and N1441) in the British North Russian Expeditionary Force.[37] *Nairana* reported that the ship had left England on 6 July 1918 and arrived at Murmansk on the 10th. A few flights were made at Murmansk at the request of the GOC Allied Forces for the purpose of "Impressing the natives." The ship then proceeded to Kem where some reconnaissance flights were

Above: Campania at Newhaven. No. 242 Squadron was formed at Newhaven with Nos. 408, 409 and 514 Flights that were equipped with Short 184 and Fairey Campania Seaplanes to cover the English Channel, and D.H.6 landplanes for short-range coastal patrols. Some seaplanes carried a "N" on their fin, fuselage and/or floats. (via H.R. Alderson)

Above: This Campania appears to have a "C" marked on top of the rear fuselage. Note how many personnel were required to launch and retrieve these floatplanes.

X.L.B

Above: Fixing the tail of a Fairey Campania into the collapsible guide on HMS *Campania*, Turnhouse, 5 July 1918. Vice Adm Richard Bell Davies, VC, DSO, AFC, served on HMS *Campania* and recalled: *The tail float had a fitting which slid into a split-tube guide set at a height which kept the machine in her flying attitude.*

Facing Page: HMS *Pegasus* under way with a Fairey Campania on the rear. The seaplane has "X.L.8" on the fuselage. Note the dazzle painting of the aircraft carrier.

made, propaganda leaflets dropped and the Kela-Petrosavodsk railway line was patrolled for signs of the enemy. While the machines were fired on by rifles, no concentrations of enemy troops were sighted.

Following the arrival of Allied forces at Archangel on 1 August, all machines were employed in attacking Modyugski Island, at the Northern Dvina River, with bombs and gunfire. Four of *Nairana's* seaplanes *then proceeded over battery & dropped bombs on conclusion of which enemy ceased fire. Troops were then landed and island captured.* The island's batteries were eventually silenced by the bombing and gunfire from the ships. *Attempts to escape from island were made in tugs. These, however, were stopped & turned by Lewis Gun fire from seaplanes.*[38] The aircraft then preceded the ships up to Archangel, dropping pamphlets on the various settlements they passed over on their way to Archangel. The Bolshevik forces had left Archangel on the night of 1/2 August, allowing the Allies to occupy it on the 2nd. *Later it was found that three river boats had got away with some of the Bolsheviks and most of the money from the banks in Archangel amounting to about forty million roubles.*

The five seaplanes had carried out 50 hours flying on the two days. Constant flying was then done from this time to the 15th, the seaplanes assisting the infantry by dropping several tons of bombs on enemy positions and the railway line such as that of Friday the 6th when in the afternoon *bombs and machine guns were used with good effect along the railway line in this district and a photograph taken, with little or no opposition. Again in the evening, a search was made in the same location, and Bolsheviks leaving in transport for the south were forced to take to adjacent woods as a result of our attack – notification of this was conveyed to 'Nairana' by W/T machine N2370.*[39] When the ship left Archangel on the 13th, the enemy held a position at Oboserskaya, about 90 miles down the railway line. They held this position for nearly a week. Kem was again visited on the 15th where the ship remained for nearly a week. Flights were made inland to find and bomb enemy positions and to patrol the railway line. N1847 with crew of Lt Thomson[40] and Lt W.J Umpleby, was brought down on the 17th by a bullet severing the petrol pipe. The pilot alighted successfully on the River Vuig, a small river near Olyupi. It proved impossible to take-off from here and as the location was about 50 miles inland in enemy territory, and the country was of such a nature that made transport practically impossible, salvage was impossible. *Information having been received that the enemy were attempting to salve the machine brought down in River Vuig, two machines were dispatched, one with view to stopping this manoeuvre, and*

Above: F.17 Campania N2366 under tow at Dundee, 3 January 1918. N2366 was fortunate in having a 250-hp Rolls-Royce Eagle VIII. These ceased at N2374, the rest of the batch of 40 (N2360–N2399) received the 260-hp Sunbeam Maori II engines. N2365 served at Cherbourg. Note the two-blade airscrew.

Above: HMS *Campania* undergoing its second conversion with the take-off deck extension.

Above: Another view of HMS *Campania* after its second conversion.

Above: Type F.22 Campania cockpit. Note the compass situated behind the windscreen and the light mounted over it.

Above: This Type F.22 Campania appears to be in a spot of bother. Even with a man on the port wing, the starboard float is basically totally submerged. Location unknown.

the other to destroy machine.[41] The seaplane was destroyed the next day by gunfire from N1844 and N1848 using incendiary ammunition and photographs taken of the burning wreck. The pilot and observer had swum to shore and successfully returned to the ship in the afternoon of the 18th.

On the 28th two machines were sent out to reconnoitre the district. *Having reached the northern end of Lake Vuig, they circled over the village of Nadvoitskaya, but nothing of military importance could be seen. Observations were then carried out over Onda, where we were fired upon by a train proceeding south, both machines being hit. As a result of this all bombs were dropped and every round of ammunition was fired at enemy, the train being brought to a standstill and good results obtained generally.*

The ship continued to move; reconnaissance flights being carried out along the coast. The country surrounding Kundalaksha was reconnoitred and Alakuri was bombed by order of the GOC Allied Forces. On 1 September the weather had improved enough to allow a flight to destroy machine gun emplacements reported by the Army. *Two 100 lb and five 20 lb bombs were dropped o hill at Koshekaia with good results, large columns of smoke being observed to rise, apparently from burning stores. Two 100 lb and one 20 lb bomb were also dropped by the second machine on these emplacements. No sign of hostility observed by either machine.* Later in the day machine emplacements on another hill were bombed.[42] Orders were then received to proceed to Archangel immediately for stores and water. Nairana left at 5.45 am on the 2nd and reached its destination at 6.30 pm.

Some aerial reconnaissance was carried out from Kum when large bodies of German troops were reported. But could not be located due to fog No flying was done until the 17th when the line from Kum was reconnoitred and some enemy trains bombed. On 20 September a diplomatic mission left in a seaplane for Solovetski *in order to endeavour to induce the monks to loan a portion of their considerable wealth to the Archangel Govt. The mission however, was a failure. Owing to bad weather, the machine could not be flown back, but was picked up by ship the following day.*[43] This concluded all flying of importance. *Nairana* arrived back in the UK on 8 October.[44]

The Campania floatplanes disappeared from RAF service a short time later, being declared obsolete in August 1919, but they had outlived their namesake. The *Campania* had been relegated to a seaplane training and balloon depot ship due to defective machinery. On 5 November 1918, the *Campania* was anchored in the Firth of Forth when a sudden, powerful squall came up causing the ship to drag her anchor. She hit HMS *Royal Oak*, the bow of the battleship breaching *Campania's* hull. She then hit HMS *Glorious*. The engine room flooded and she lost all electrical power. She took five hours and sank by the stern. All crew were rescued. The ship is still where she sank

Above: Hoisting a salvaged Type F.22 Campania.

Serial Allocation			
Serial	**Contract No.**	**Manufacturer**	**Notes**
N1000–N1009	C.P.1115430/16 & C.P.116827	Fairey	RR Eagle variations. N1006 Sunbeam Maori II.
N1840–N1889	A.S.3591	Barclay, Curle	N1852-N1889 cancelled.
N1890–N1959		Sunbeam & Fredrick Sage	Breakup of contract not known. Cancelled.
N2360–N2399	A.S.18939	Fairey	From N2374 Maori II
N2580–N2629	B.R.177	Robey & Co	Cancelled.
HMS *Campania* had the following Fairey Campania seaplanes at various times: N1001, N1002, N1003, N1005, N1007, N1008. N1009, N1842, N1843, N1845, N1850, N1851, N2360, N2361, N2362, N2364, N2366, N2372 and N2373.			

together with the four Campania seaplanes and seven Sopwith 1½ Struters she had on board when she sank. The site is a Historic Marine protected Area.[45]

The *Nairana* had a better fate. After service in Russia the ship was sold and rebuilt to its original plans. It served on the Australian Bass Straight run for 26 years from 1921. She did not serve in World War II but was equipped with a 4-in anti-submarine and a 20-mm anti-aircraft gun and paravanes. She was the only Bass Straight ferry to operate during the war years. Laid up in 1948 she was driven ashore in a gale in 1951, and scrapped where she lay during 1953–1954.

Endnotes

1. Report of D Section of the Air Dept of the Admiralty for W/E 27.10.1916. TNA AIR1/337/15/226/247.
2. The IWM website has a film that shows N1000 under test. https://www.iwm.org.uk/collections/item/object/1060023148.
3. J.M. Bruce noted that despite the claim made in *Jane's All the World's Aircraft 1919,* no Fairey Campania used the Fairey Patent Camber Gear. Bruce, J.M. "Fairey Campania", *Air Pictorial*, May 1963, P.142.
4. 10th Weekly Report of The Controller, Technical Department, 25.04.17. TNA AIR1/2094/207/12/10.
5. TNA AIR1/6A/4/48.

Operation of HMS *Nairana's* Aircraft in Russia
16 August to 27 September 1918

Date	Machine	Pilot	Observer	Time	Place	Remarks
16/8	N1848	Capt Simpson	Lt Hopwood	3-50	Kom	Reconn & Bombing "
"	N1847	Capt. Will	Lt Stewart	3-15	"	Reconn Segezsha.
17/8	N1848	Lt Bridge	Lt Wallace	2-45	"	Reconn Pagiatkaya & Maidesakaya
"	N1848	Capt Simpson	Lt Stewart	1-05	"	Reconn Segezha.
"	N1847	Lt Thompson	Lt Umpleby	0-45	"	M/c brought down landed in river Nr. Olympi.
18/8	N1848	Capt Simpson	Capt McNab	3-33	"	Destroyed abandoned M/c with incendiary gun fire.
"	N1844	Capt Will	Lt Stewart	3-20	"	"
20/8	N1848	Capt Will	Lt Stewart	2-55	Soroka	Reconn Lake Vuig
"	N2370	Capt Halsted	Capt Malet	2-50	"	Reconn Nadvaitakaya & Lake Vuig.
21/8	N1849	Lt Thomson	Lt Umpleby	2-45	"	Reconn Pevyenes, Segezha replied to hostile rifle fire with MG.
"	N1849	Capt Simpson	Lt Wallace	2-05	"	Reconn up River Vuig.
23/8	N1849	Capt Fryer-Smith	Lt Wright	2-35	Kandalaksha	Reconn Alakurti.
24/8	N1844	Lt Bridge	Lt Hopewood	3-50	"	"
"	N1849	Capt Simpson	Capt Malet	2-00	"	Bombed Alakurti & trenches.
25/8	N1849	Capt Will	Lt Stewart	2-55	"	Reconn Alakurti, bombed village & trenches.
26/8	N1849	Capt Fryer-Smith	Lt Wright	3-50	"	Reconn Alakurti.
28/8	N1848	Capt Will	Lt Stewart	2-33	Soroka	Reconn Nadvaitakaya & Onds. Bombed & MG hostile train.
"	N1849	Capt Simpson	Capt McNab	2-30	"	"
1/9	N2370	Capt Halsted	Capt Malot	2-51	Onega	Bombed & MG enemy gun emplacements & stores dump.
"	N1844	Capt Fryer-Smith	Lt Earle	1-53	"	"
"	N1848	Lt Thompson	Lt Umpley	1-20	"	"
3/9	N1848	Capt Will	Lt Stewart	3-50	Archangel	Bombed MGnd & dropped pamphlets on enemy positions at Oboserakaya.
"	N1844	Lt Bridge	Lt Wallace	3-20	"	"
4/9	N1848	Capt Simpson	Capt McNab	3-35	Kom	Reconn Uktinskaya.
"	N1844	Capt Fryer-Smith	Lt Wright	3-25	"	"
"	N1844	Capt Will	Lt Stewart	2-55	"	"
5/9	N1848	Capt Simpson	Capt McNab	4-00	"	Reconn Perveakaya.
16/9	N2370	Capt Halsted	Lt Earle	0-25	Kandalaksha	Local observation.
18/9	N1844	Capt Fryer-Smith	Lt Wright	3-00	Kom	Reconn Segezha. Bombed train.
"	N1849	Lt Bridge	Lt Wallace	3-20	"	
"	N1848	Capt Simpson	Capt McNab	3-00	"	Reconn Nadvaitakaya Bombed enemy firing MGs from houses.
23/9	N2370	Capt Simpson	Capt Mcnab (sic)	2-35	"	Reconn up Kam River.
27/9	N1844	Capt Will	Capt Macnab(sic)	0-35	Kandalaksha	Local observation.

Source: TNA Air1/292/15/226/140.

Above & Below: N2377 with 26-hp Maori II engine. The Maori engines were introduced at N2375. This seaplane served with 410 Flight at Calshot from March 1918, and also at Bambridge in September 1918. Deleted W/E 3 October 1918.

Above, Below, & Facing Page, Top: Sunbeam Maori II powered F.22 Campania N2392 fitted with the Connor Silencer, No. 242 Squadron, 15 August 1918. Note the "N" on the fuselage for "Newhaven". The Connor silencer was one of five tried out in 1918-1919. In the Connor the exhaust gas was passed from the inlet pipe through an annular space between the pipe and an outer one of 7 in diameter. The gases are guided along this annular path by means of three turns of an expanding spiral of metal. While lighter than other silencers, at 24½ lbs, it was more complicated in construction and harder to clean effectively. The *Report on Silencer Experiments Carried out at Butley A.E.S (1918–19)* mentions the Connor being tried out in the Bristol Fighter, de Havilland D.H.9 and the F.E.2b but no mention of any seaplanes. (Source: TNA AIR1/2094/207/12/8).

Above: N2362 on the sling, HMS *Campania*. The meaning of the large fuselage letters is presently unknown. Note the bomb carrier under the fuselage.

Above: Judging by the fuselage markings, this is probably N2362 taking off from HMS *Campania*. The wheels fell off when the machine left the deck. This photograph was one of a trio showing the launching of a seaplane from the *Campania's* deck in Richard Bell Davies autobiography *Sailor in the Air.*

Calshot – Aircraft Strength at End of War – 1918

Floatplanes Type	240 Squadron Nos. 410/411 Flights	210 T.D.S. B Flight	Calshot Group Workshop & Repair Depot
Campania	N2376, N2395	N2381, N2386, N2387, N2389, N2393	N2379, N2399
Short 184 (Maori)	N2985, N9108	N1623, N2951, N2952	N2975
Short 320 (Cossack)	N1707, N1709	N1403	N1366, N1398, N1402
Wright S.P. (Maori)	-	N1280, N1282, N1283	N2185
Wright S.P. (Renault)	-	-	N1281

Type	240 Squadron No. 346 Flight	210 T.D.S. A Flight	Calshot Group Workshop & Repair Depot
Felixstowe F.2A	N4081, N4285, N4286, N4535	N4292, N4534, N4547	N4309, N4536, N4544
Curtiss H-12	N4334, N4337, N4349		

Policy changed for Calshot after the war and by early 1919 only 3 F.1A boats and 2 Short 184 floatplanes were on Group operations Strength and the Training School has become a Floatplane Training School. Apart from a single F.2A (N4309) all other aircraft were Maori powered Short 184 floatplanes: N1621, N1623, N2290, N2999, N9039, N9042, N9050, N9091, N9110, N9111, N9174, N9175, N9176 and N918.

Above: Sunbeam powered Fairey Campania floatplane as indicated by the central exhaust stack.

Right: If the Campania had faced a single Hansa-Brandenburg W.29, let alone three W.29 monoplanes, the result would have been fatal for the Campania. The Campania did not take part in any spectacular action in WWI and its work in Russia was overlooked such that it did not get much coverage in the pulp magazines between the wars. This issue from January 1932 of *Model Airplane News* is one of the few times it graced a cover.

Specifications				
Source	**1**	**1**	**1**	**2**
Type	**F.16**	**F.17**	**F.22**	**N1006**
Dimensions				
Span upper	61 ft 7½ in	61 ft 7½ in	61 ft 7½ in	
Folded				
Lower				
Chord	6 ft 4 in	6 ft 4 in	6 ft 4 in	
Length	43 ft 35/8 in	43 ft 05/8 in	43 ft 05/8 in	
Folded				
Height	15 ft 1 in	15 ft 1 in	15 ft 1 in	
Gap	6 ft 6 in	6 ft 4 in	6 ft 4 in	
Width across floats				
Incidence				
Dihedral				
Span tail	13 ft	13 ft	13 ft	
Incidence Tailplane				
275 RR side radiators				
275 RR front radiator				
Eagle VIII				
Sunbeam Maori				
Areas				
Wings in sq ft	639.8	627.8	627.8	655
Total Tail (including elevators)	75.1	75.1	75.1	
Elevators	. 28.1	28.1	28.1	
Rudder	22	19.6	19.6	
Fin	14	16.2	16.2	
Aileron	23.4	23.4	23.4	

Notes: 1. This fuel load gave 4½ hours endurance at full speed at sea level.
2. To this must be added about 100 lbs of water that was found in the floats at the conclusion of the trials. With 135 gallons and no bombs the machine had a useful range of 456 sea miles.
3. Does not include fuel.

Source:

1. *Jane's All the World's Aircraft 1919.*
2. TNA AIR2/969. Trials were conducted at Grain on 18 & 19.10.17. N1006 had a fixed tailplane and a single Lewis gun on Scarff ring with three trays of ammunition
3. Technical Notes Fairey Campania (Type F.22).
4. *Types of British Seaplanes.* August 1918. Copy in TNA AIR10/96.
5. Minutes of 5th Air Department Weekly Progress Meeting of 12.06.1917. TNA AIR1/270/15/226/113.
6. Performance report. Papers of W Cdr N.H. Woodhead, commander Marine Acceptance Branch, RNAS Dundee Station. Test probably that of 7 May 1918. *Cross & Cockade Great Britain*, Vol.9 No. 2 1978.

Specifications

Source	3	4	5	6
Type	F.22	Campania	Campania	N1842
Dimensions				
Span upper	61 ft 7½ in	61 ft 8 in		61 ft 7½ in
Folded				
Lower	45 ft 11 in			
Chord	6 ft 4 in			
Length	43 ft 0 in	43 ft 3 in		
Folded	43 ft 9 in			
Height	14 ft 7 in	15 ft 1 in		
Gap	6 ft 6 in			
Width across floats				
Incidence	5°			5°
Dihedral	2°			2°
Span tail				
Incidence Tailplane				
275 RR side radiators	4°			
275 RR front radiator	3°			
Eagle VIII	2°			2½°
Sunbeam Maori	¾°			
Areas				
Wings in sq ft				
Total Tail (including elevators)				
Elevators				
Rudder				
Fin				
Aileron				

6. The story of the Air Department A.D. Flying Boat is told in Owers, C.A. *AD Flying Boats,* Albatros Productions, Datafile No. No. 159, UK. The wartime constructed A.D. boats were not a success and did not perform the tasks that the Campanias undertook, being mainly relegated to training. They were modified into a successful flying boat post-war by Supermarine Aviation.
7. Experimental machines received & under test. 14th Weekly Report of the Controller, Technical Dept for W/E 23.05.1917. Copy in RAF Museum J.M. Bruce Collection Box 25.
8. 14th Weekly Report of the Controller, Technical Department for W/E 23.05.1917. Copy in RAF Museum J.M. Bruce Collection Box 25.
9. *Technical Notes for the Fairey Campania (Type F.22),* RAF, 1918. Copy in TNA AIR 10/617.
10. From notes on a meeting held 26.04.1917 contained in a CTD Report f 23.05.1917. Photocopy Report in RAFM JM Bruce Collection Box 25.
11. Rear Adm Sims to Sec Navy J Daniels. 12.04.1917. USN History and Heritage Command.
12. Later S/Ldr M Wright became a Director of Fairey Aviation Co. Flying from the Isle of Grain to Scapa Flow was, for the time, a considerable achievement for a seaplane and duly promoted by Fairey. See Smith, A. *The Man Who Built The Swordfish* for a view of how this flight strengthened the symbiotic relationship between Fairey and the Admiralty's Air Department.
13. Arthur Vyell Vyvyan joined the Royal Navy in 1891. He was promoted to Lieutenant in January 1896. In 1902 he was posted to the Sheerness Gunnery School senior staff on HMS *Wildfire.* In 1916, Captain became involved in

Specifications				
Source	**1**	**1**	**1**	**2**
Type	**F.16**	**F.17**	**F.22**	**N1006**
Weights in lbs				
Disposable load	1,026	1,1003	1,0263	52
Lewis Gun & Ammunition				23
Scarff ring				
Bomb armament				
Gun armament				450
4 x 100-lb bombs & gear				116
W/T				
W/T & Signalling lamp				25
Sundries				360
Crew				
Water				3,762
Empty				3,762 (12 gall water)
Fuel				555 (78 gal)
Oil				76
Fully loaded	5,500	5,560	5,329	5,329
Capacities in gall				
Fuel	88	88		78
Oil				8
Climb to				
1,000 ft				3 min 15
2,000 ft				7 min 00c
3,000 ft				11 min 15
4,000 ft				16 min 40
5,000 ft	14 min	12.30 min	18 min	23 min 30
6,000 ft				32 min 15
6,500 ft				38 min 00
7,000 ft				45 min 30
8,000 ft				
10,000 ft				

the RNAS, serving as the Assistant to the Superintendent of Aircraft Construction before being appointed the Assistant to the Director of the Air Service. Departing London just after the 1 April 1918, Vyvyan briefly returned to operational service as the senior RAF officer in the Mediterranean. In late April he was recalled to Great Britain and appointed the General Officer Commanding the RAF's Norther Area. The following year he became Air Officer Commanding Coastal Area, being promoted to AVM shortly after taking up the post. He retired from the RAF in 1925, as Sir Arthur Vyvyan, KCB, DSO.

14. Minutes of 4th Air Department Weekly Progress Meeting held on 05.06.1917. TNA AIR AIR1/270/15/226/113.
15. Minutes of 5th Air Department Weekly Progress Meeting held on 12.06.1917. TNA AIR AIR1/270/15/226/113.
16. Minutes of 6th Air Department Weekly Progress Meeting held on 19.06.1917. TNA AIR AIR1/270/15/226/113.
17. *Ibid.* Also, 22nd Weekly Report of the Controller, Technical

Specifications				
Source	**3**	**4**	**5**	**6**
Type	**F.22**	**Campania**	**Campania**	**N1842**
Weights in lbs				
Disposable load				
Lewis Gun & Ammunition			75	
Scarff ring				
Bomb armament				50
Gun armament				32
4 x 100-lb bombs & gear			450	
W/T				
W/T & Signalling lamp			125	
Sundries				
Crew			360	320
Water				100
Empty			3,613	3,794
Fuel			(102 gall1)	658 (94 gall)
Oil				60 (8 gall)
Fully loaded			5,4062	5,014
Capacities in gall				
Fuel			135	
Oil				
Climb to				
1,000 ft				2 min
2,000 ft				4 min 10
3,000 ft				
4,000 ft				9 min 20
5,000 ft				
6,000 ft				14 min 40
6,500 ft				
7,000 ft				
8,000 ft			44 min	22 min 15
10,000 ft				32 min 20

Dept, for W/E 18.07.1917. Copy in RAFM J.M. Bruce Collection Box 25.

18. Fairey Aviation Co, Weekly Report. 21.11.1917. TNA AIR2/969.

19. A Smith in his book on Sir Richard Fairey states that Fairey switched between Sunbeam and Rolls-Royce engines to maximise the performance of the Campania, apparently not realising that the Rolls-Royce engines were preferred but were not available. *The Man Who Built The Swordfish – The Life of Sir Richard Fairey,* I.B. Tarus, UK, 2018, P.87.

20. Minutes of the 14th Meeting of the Air Department Weekly Progress Meeting of 14.08.1917. TNA AIR AIR1/270/15/226/113.

21. Dept Aeronautical Supplies 24th Report for W/E 18 August 1917. TNA AIR 1/500/15/331/5.

22. Minutes of the 15th Air Department Weekly Progress Meeting of 21.08.1917. TNA AIR AIR1/270/15/226/113. N1006 was tested with 200-hp Sunbeam No. 1/250/774

Specifications				
Source	**1**	**1**	**1**	**2**
Type	**F.16**	**F.17**	**F.22**	**N1006**
Speed	42 mph	43 mph	40 mph	
Landing	72 kts	78 kts	74 kts	
Low down				73½ kts
at 2,000 ft				
at 3,000 ft				
at 5,000 ft				
at 5,800 ft				68 kts
at 6,500 ft				
at 7,500 ft				
at 10,000 ft	250-hp RR	275-hp RR	250-hp Sunbeam	260-hp Sunbeam
Engine				

with a honeycomb radiator of 222 ft^2 at the Isle of Grain on 18.10.1917. TNA AIR2/969.

23. Minutes of the 16th Air Department Weekly Progress Meeting of 28.08.1917. TNA AIR AIR1/270/15/226/113.
24. Loose sheet from Report, Controller, Technical Department. RAF Museum J.M. Bruce Collection Box 64 Loose sheet.
25. Post-war the D.H.9A and Bristol Fighters, both types having frontal honeycomb radiators, operating in the desert with additional honeycomb radiators added and performed well.
26. Minutes of the 23rd Air Department Weekly Progress Meeting of 16.10.1917. TNA AIR AIR1/270/15/226/113.
27. Minutes of the 24th Air Department Weekly Progress Meeting of 23.10.1917. TNA AIR AIR1/270/15/226/113. According to Bruce Robertson 100 Campanias were ordered from Sunbeam and Sage but cancelled. Letter to JMB 08.08.1963. RAFM JM Bruce Collection Box 26.
28. Notes from 27th Report of the CA Directorate of W/E 8.09.1917. RAF Museum, J.M. Bruce Collection Box 24.
29. This installation of high lift wings is most probably that which led *Jane's All the World's Aircraft 1919*, to state that N1000 was designed with a "fairly high-lift wing section."
30. Copy in TNA AIR2/969.
31. Bruce, J.M. *British Aeroplanes 1914–1918,* Putnam, UK, 1957. P.220.
32. Copy in TNA AIR2/969.
33. Minutes of 23 Air Department Weekly Progress Meeting. 14.11.1917. TNA AIR1/270/15/226/113.
34. Minutes of 23 Air Department Weekly Progress Meeting. 18.12.1917. TNA AIR1/270/15/226/113.
35. Taylor, H.A. *Fairey Aircraft since 1915,* Putnam, UK, 1984. P.60.
36. *RNAS Operations Reports,* Reprint, Naval & Military Press, UK, 2019.
37. Philpott, I.M. (*The Royal Air Force – An Encyclopedia of the Inter-war Years Vol I,* Pen & Sword, UK, 2005) has incorrectly shown *Nairana's* Campania floatplanes being involved in Russia in 1919, whereas they were employed in 1918, Short 184 and Fairey IIIC seaplanes were involved in RAF actions in 1919.
38. Nairana. Hand written log of air operations. AIR1/435/15/274/2.
39. Nairana. Hand written log of air operations. *This log is not intended to be official but was kept merely as a diary for reference.* Signed Halsted. AIR1/435/15/274/2.
40. Spelling as in original report. TNA Air1/292/15/226/140.
41. Nairana. Hand written log of air operations. AIR1/435/15/274/2.
42. Nairana. Hand written log of air operations. AIR1/435/15/274/2.
43. Nairana. Hand written log of air operations. AIR1/435/15/274/2.
44. This section has been written from TNA Air1/292/15/226/140.
45. https://en.wikipedia.org/wiki/HMS_Campania_(1914). Sturtivant & Page do not list any *Campania* floatplanes lost in the sinking of the ship.

Specifications				
Source	**1**	**1**	**1**	**2**
Type	**F.16**	**F.17**	**F.22**	**N1006**
Speed				
Landing				
Low down				74 kts
at 2,000 ft				
at 3,000 ft			76 kts	73 kts
at 5,000 ft				68 kts
at 5,800 ft			74 kts	
at 6,500 ft				68 kts
at 7,500 ft			66 kts	
at 10,000 ft				63 kts
Engine	RR Eagle VIII	275-hp RR	275-hp RR	345-hp RR Eagle 8

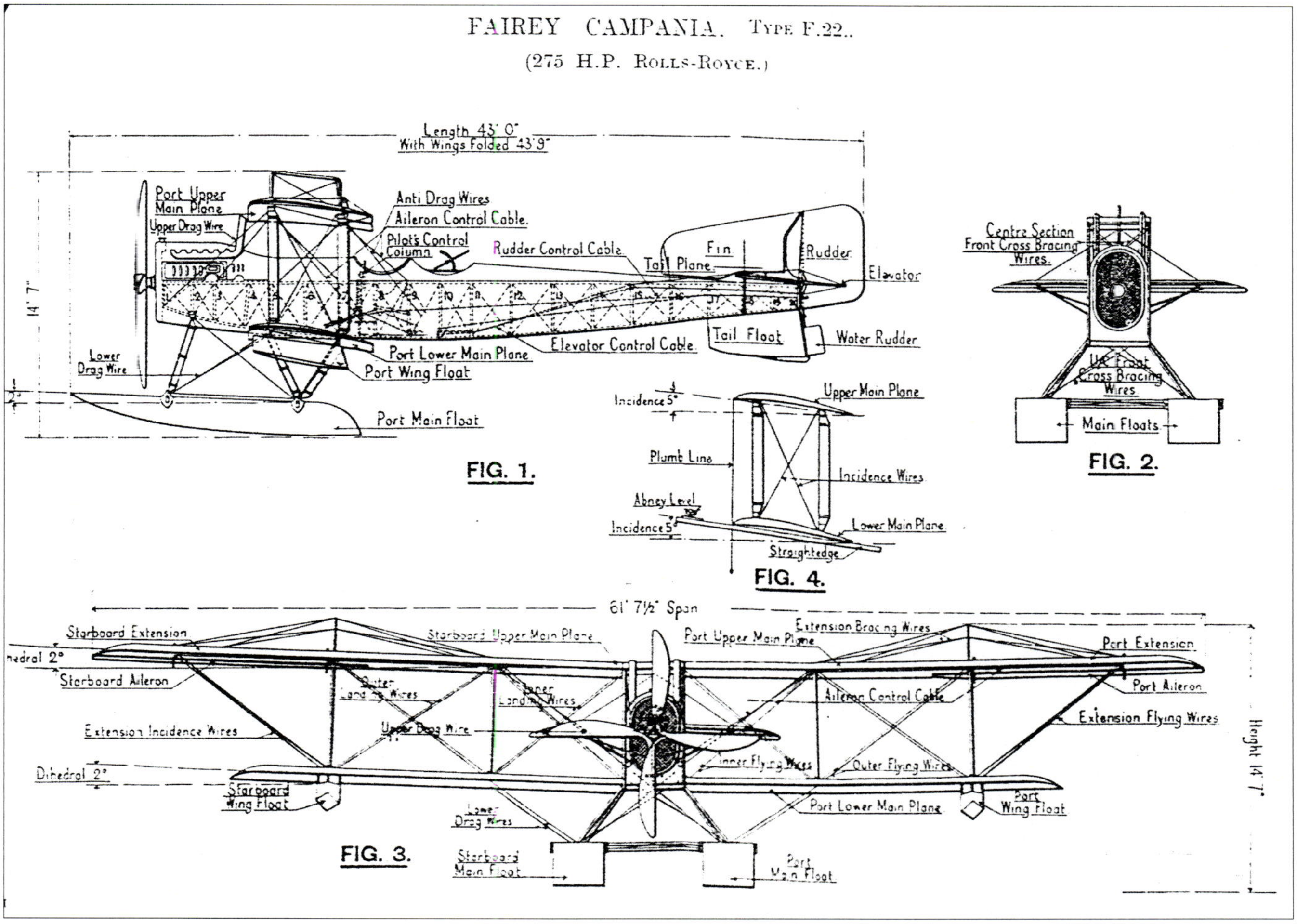

Campania

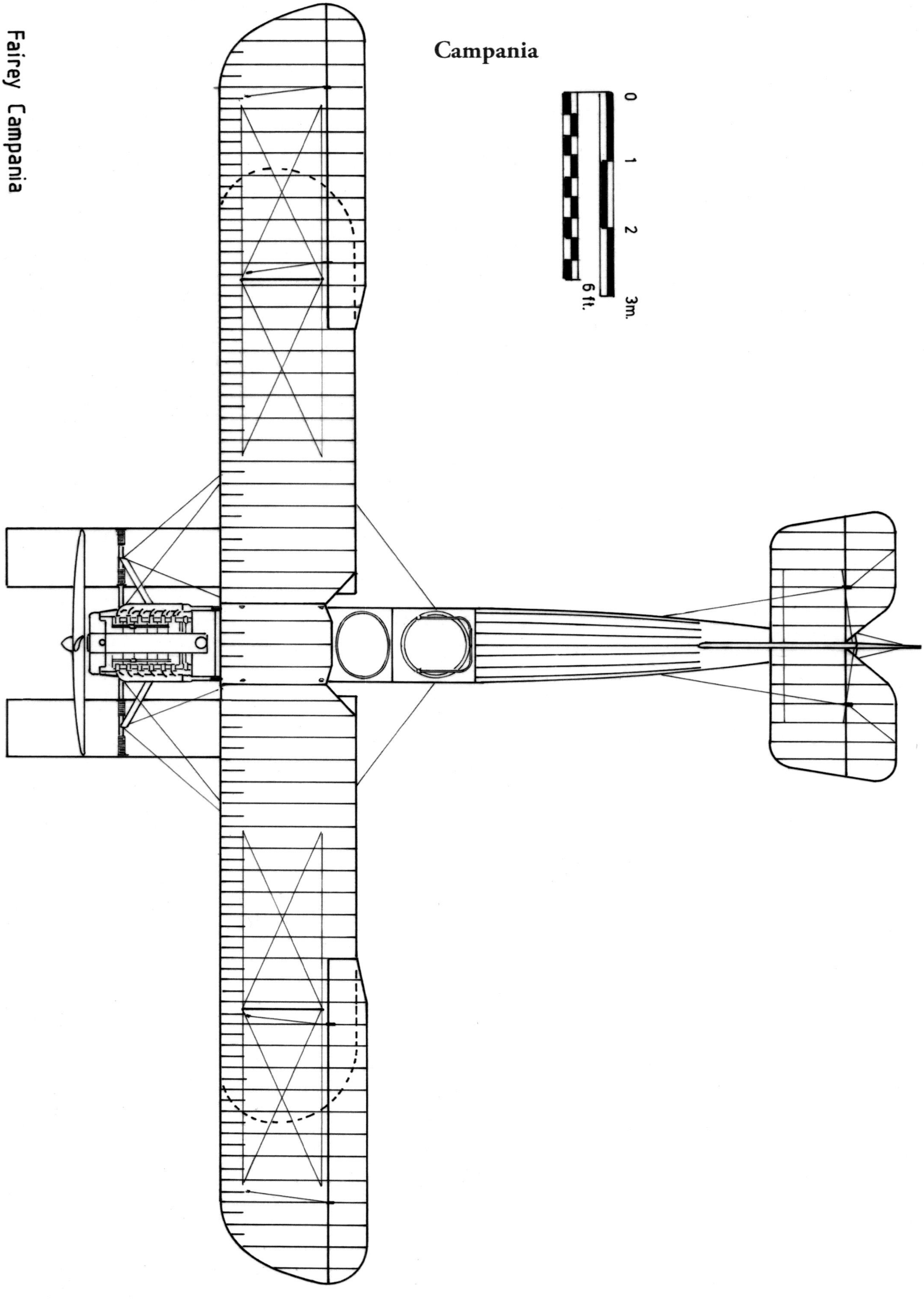

Fairey Campania

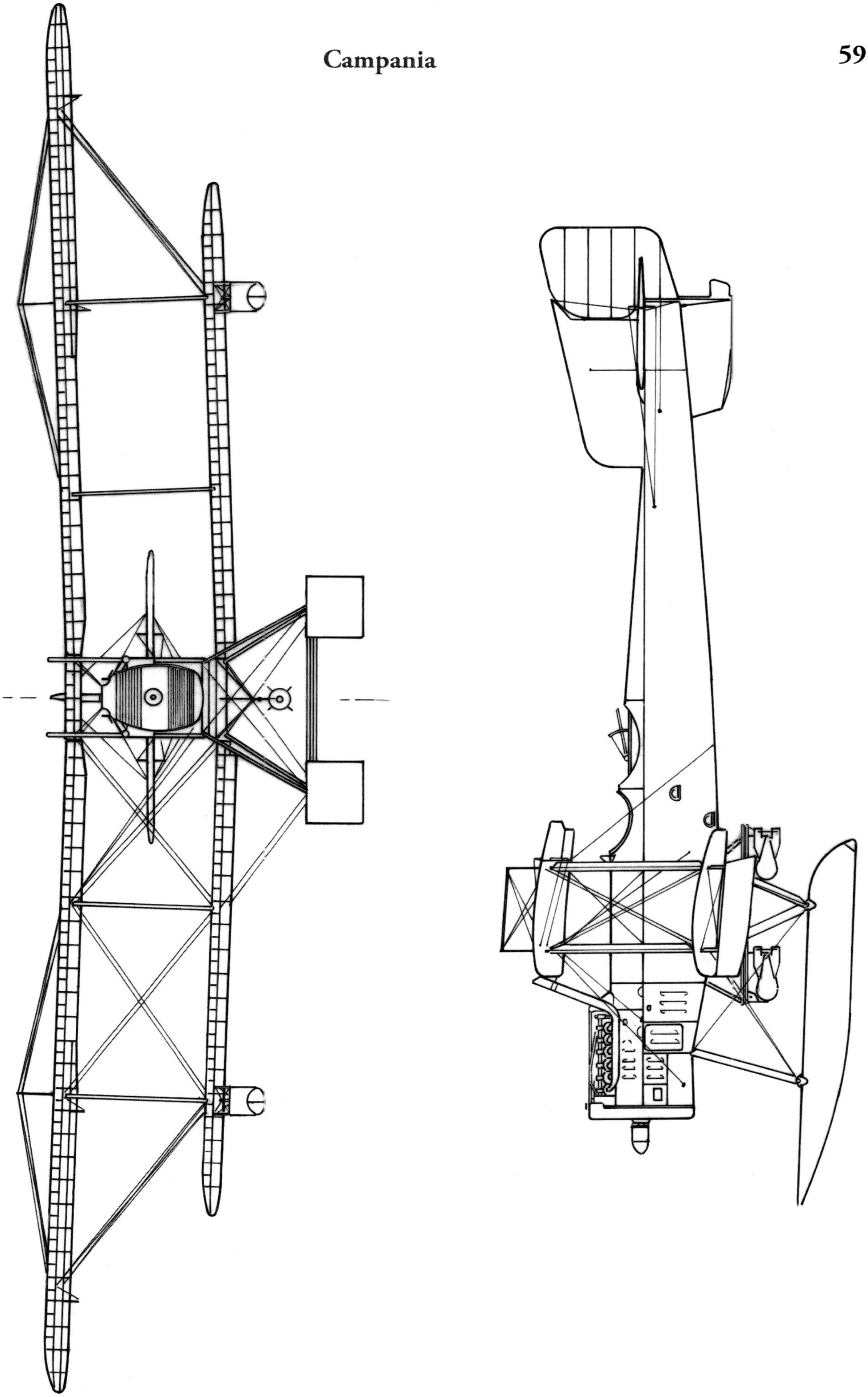

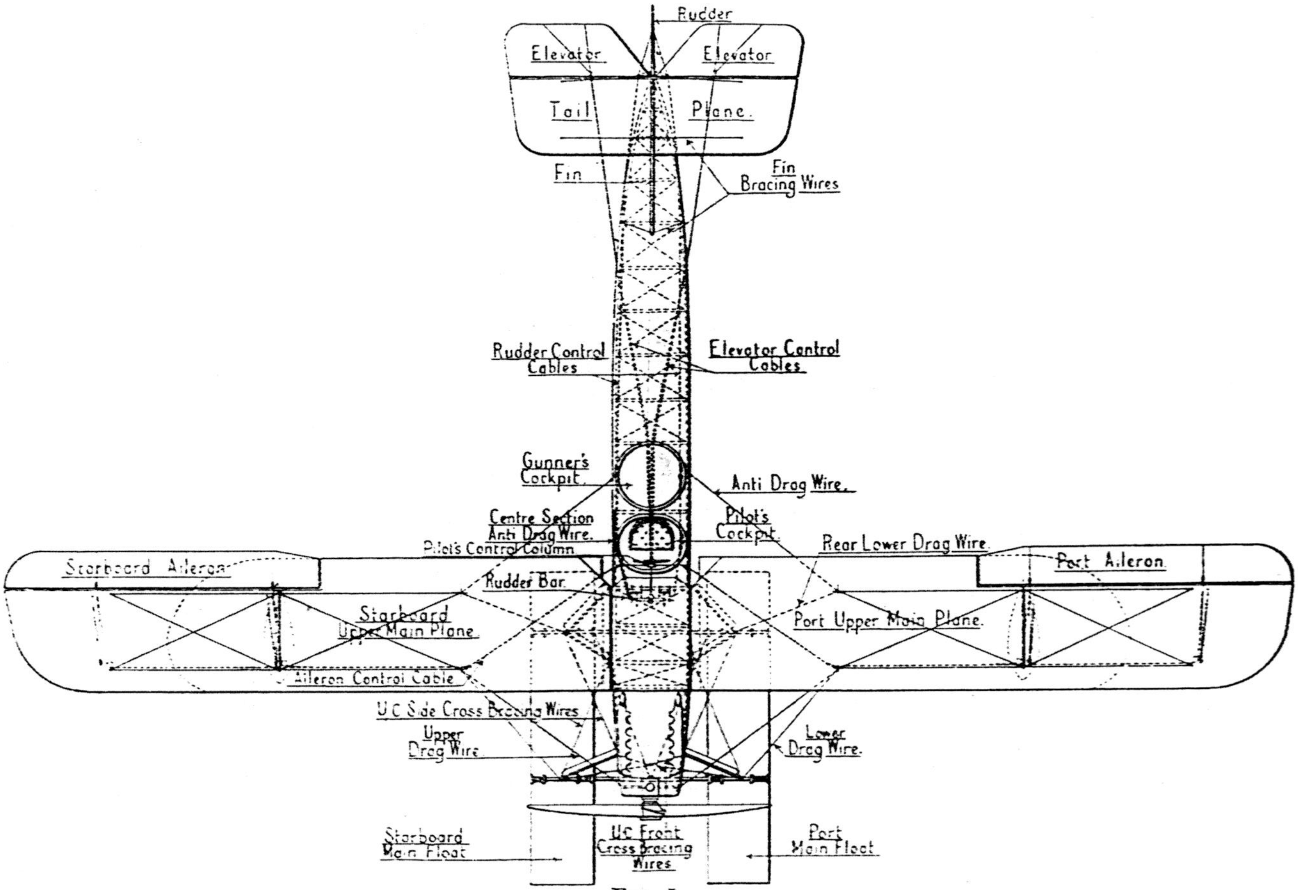
Rudder
Elevator
Elevator
Tail
Plane.
Fin
Fin
Bracing Wires
Rudder Control
Cables
Elevator Control
Cables
Gunner's
Cockpit.
Anti Drag Wire.
Centre Section
Anti Drag Wire.
Pilot's
Cockpit.
Pilot's Control Column
Rear Lower Drag Wire
Starboard Aileron
Port Aileron
Rudder Bar
Starboard
Upper Main Plane
Port Upper Main Plane.
Aileron Control Cable
U.C. Side Cross Bracing Wires
Upper
Drag Wire
Lower
Drag Wire.
Starboard
Main Float
U.C. Front
Cross Bracing
Wires
Port
Main Float

Fig. 5.

Above: Main float shock absorber.

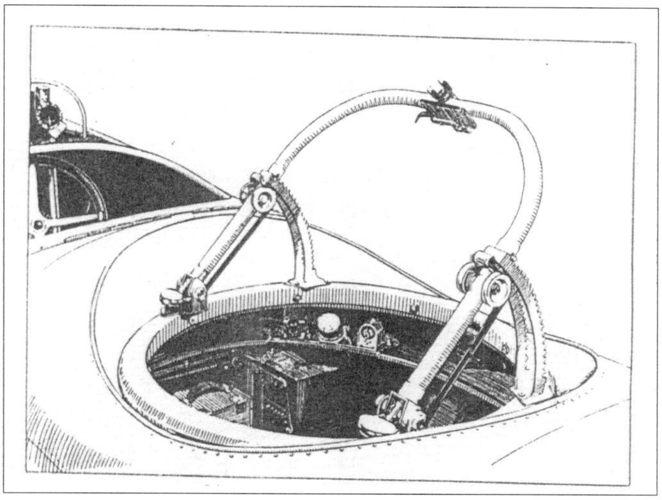

Above: Gun ring detail.

Below: Tail plane and rear float detail.

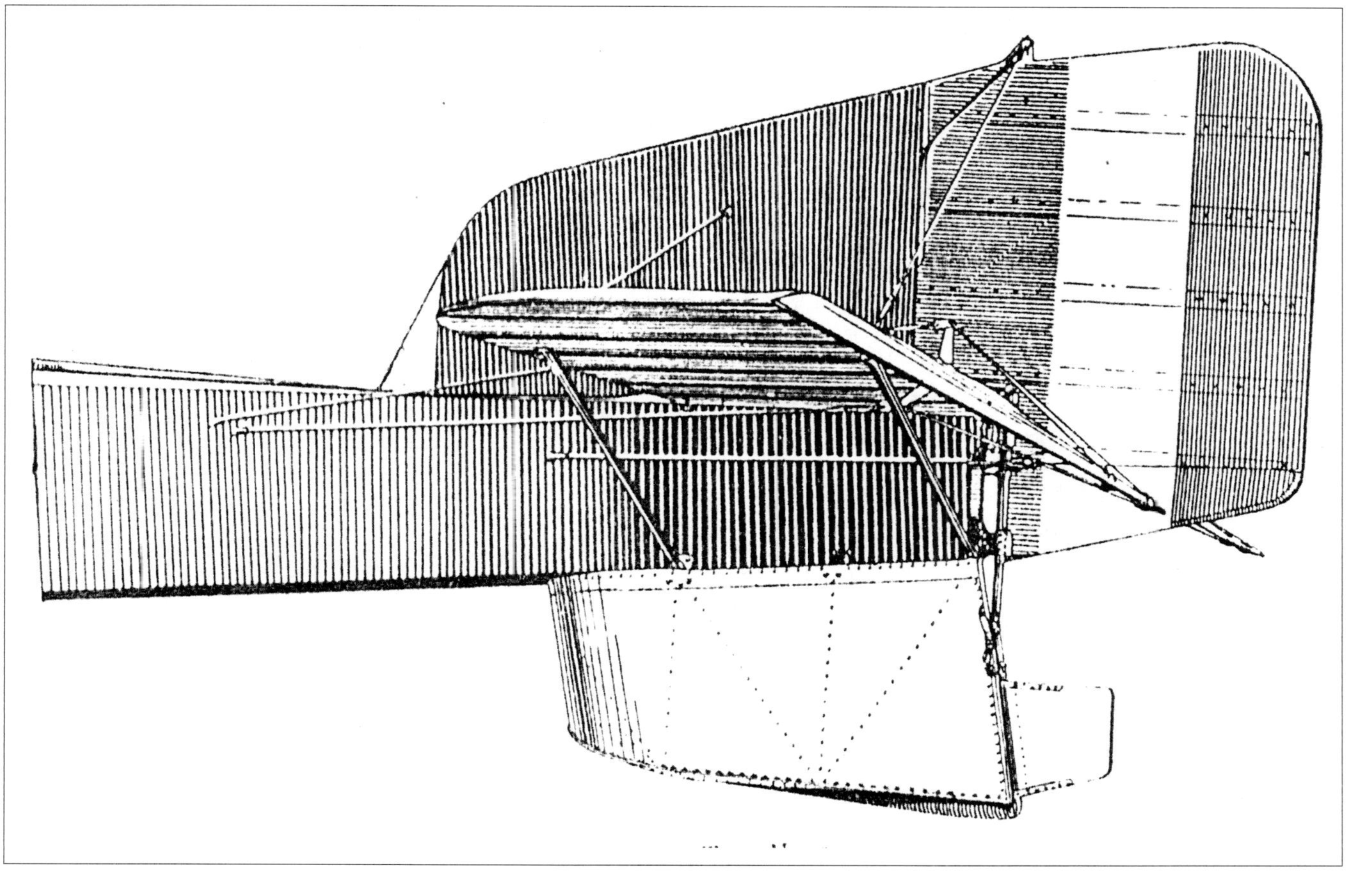

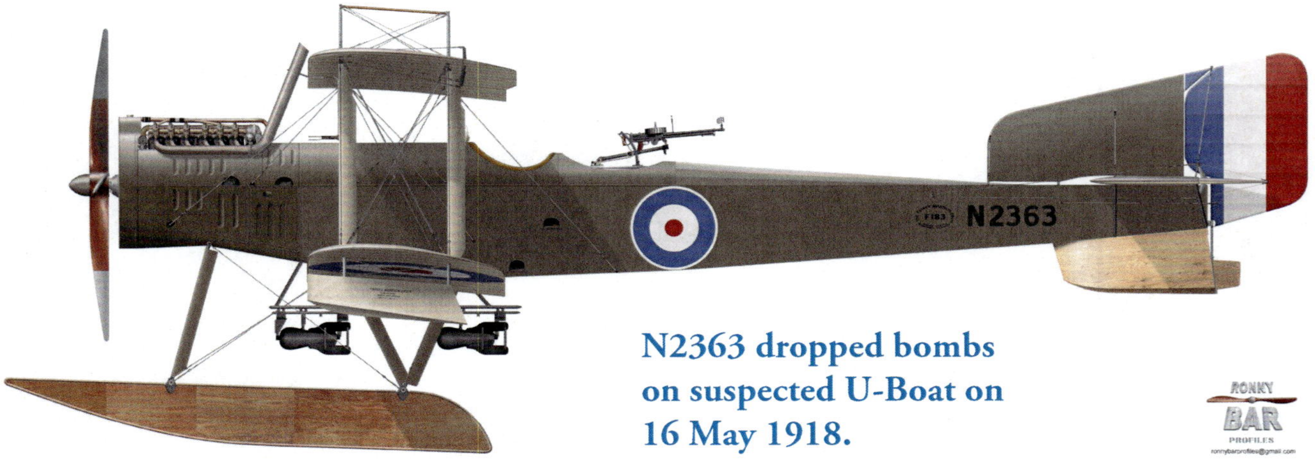

N2363 dropped bombs on suspected U-Boat on 16 May 1918.

N2368 bears white codes on the fuselage.

N2392 with the Connor silencer fitted. Note the "N" for Newhaven on the fuselage.

Chapter 4: The Fairey Hamble Baby

Above: Sopwith Baby No. 8134 in Fairey's workshop after conversion to a Hamble Baby.

At 0800 on 23 October 1917, Flt. S/Lt. Lander left Westgate in Seaplane No. N.1203 on a special patrol in search of a reported submarine. After flying about 30 minutes the pilot sighted a periscope on his starboard bow in position 25-30miles East of N. Foreland. The periscope was causing a continuous wake which was noticeable in the broken sea, and on the seaplane approaching nearer, a disturbance was seen in the water behind it. The periscope then submerged, and the machine passed over the disturbance from the stern, dropping two 65-lb bombs, from about 800–1,000 feet, which exploded 100 feet in front of the wash. The area was searched for twenty minutes and large quantities of oil came to the surface. The sea was too rough to discern any bubbles. Owing to the wind changing and increasing, the pilot lost his bearings and finally landed at Calshot. Throughout the patrol the visibility was poor, with continuous low clouds.[1]

Lander's aircraft was a single seat Fairey Hamble Baby floatplane. Sopwith had developed the successful Baby fighter seaplane from the 1914 Schneider floatplane. By 1916, the Sopwith Baby was being flown with loads that meant that it could only get airborne in none but the most favourable wind and sea conditions. The load had grown to include two 65-lb bombs for anti-submarine patrols, a synchronised Lewis gun, its ammunition, sea anchor and pigeons (to be released with a message if forced down). The floats were often broken by the pounding of the waves while the biplane attempted to take-off. While power could be increased by installing a 130-hp Clerget instead of the 100-hp Gnôme Monosoupape rotary engine, the lifting power of the flat, thin aerofoil sections then in use, could be improved by using a more heavily cambered section such as were explored at the National Physical Laboratory. Blackburn, who had taken over the development of the Baby from Sopwith

Above: Another view of Sopwith Baby No. 8134 in Fairey's workshop after conversion to a Hamble Baby.

due to that firm's workload, adopted a high-lift section for its modification of the Baby such that were often termed Blackburn Babies in official documents.

Sqn Cdr John W Seddon of the Experimental Construction Department of the Marine Experimental Depot based at the Isle of Grain, had been working on the problem. He had produced the Port Victoria P.V.1 by drastically modifying a Baby seaplane and it had lifted some 400 lbs more than the standard Baby although it weighed 300 lbs more than an unloaded Baby.

Calshot Confidential Memorandum of 5 May 1917

The following report was based on the experience of the officers attached to the Submarine Patrol based at Calshot.

It has been found necessary to divide the patrol into two distinct types, ie., the Standard and the Special patrol. The Standard patrols consist of definite areas, over which the seaplane patrols. The Special patrol is a patrol carried out direct to a given position where a submarine has been reported or is suspected.

Both Sub Station in the area has its own standard patrols each one being a definite area enclosed by a triangle or rectangle with fixed extremities and covering a part of the Transport or Trade routes.

Each Sub Station carries out a number of patrols over its own standard areas each day. If a Transport patrol, then at Daybreak and in the Evening, when the Transports are actually crossing, or at other times if required. If a Trade Route patrol, then at irregular times, or continuously if a submarine is suspected in that area.

The Calshot Patrol covers an area combining the Trade Route Standard patrols of Portland and Bembridge, working over these areas at times between those of the Sub Stations. In this way it is possible to keep at least one machine on the Trade Route practically continuously all day.

The Special Patrols are regulated from Calshot for all Sub Stations, Calshot deciding which Sub Stations should send out a Seaplane to any position where a submarine is reported, regulating this according to the weather. This method obviates the chance of confusion or of all Sub Stations sending their Stand-by machines away together, as might happen if working independently.

Calshot also decides which Standard patrol is to be carried out, and at what times, for all Sub Stations, regulating this according to the submarine reports, amount of Shipping, weather etc.. It has been found absolutely essential, in order to maintain an efficient patrol in this area, to regulate the patrols entirely from Calshot, all submarine reports, etc. being signaled direct to Calshot from the Commander-in-Chief's Office. It is possible however, for the local S.N.O. in the vicinity of a Sub Station to send out a Special Patrol if desired, the fact being signaled to Calshot at once.

Above: This photograph is believed to show Sopwith Baby No. 8134 when it arrived at Fairey's Hamble works for conversion with high lift wings. The shape of the thin aerofoil sections can be seen. The trolley for No. 8559 was for a Fairey manufactured Short 827. Note how far the water is from the shingle beach.

Above: A production Hamble Baby at Calshot, shows the difference in the wings from the Sopwith Baby. Note Curtiss H-12 flying boat in water in background.

Above: Parnall Hamble Baby No. N1970 appears to be at the same location and possibly the same machine. No. N1970 was at Calshot from November 1917. Deleted W/E 29 August 1918. The significance of the "P.1." under the serial is unknown.

In the W/E 14 April 1916, Y Section of the Admiralty Air Department recorded the decision to place an order with Fairey for six sets of experimental planes for the Sopwith Baby. The following week Sqn Cdr Seddon's visit to the Fairey Company with respect to "Special sections" for Baby seaplanes was recorded.[2] Each set of wings was to be of a different aerofoil section. The Fairey Aviation Co was given a standard Sopwith built Baby, No. 8134, for the experimental development work.

Baby seaplane 8134 Experimental Wings was reported in late June as at Hamble and exhibiting *very good speed range and climb but requires new tail float before trials with full war load.* These trials were stated to be made around 8 July. The pilot was S.C. Seddon. The report concluded with the note that the *No.2 experimental wings probably ready for testing about 15 July.*[3]

Fairey adopted a completely different approach with their second set of wings. Fairey conceived the idea of deflecting the entire trailing edge of the wing but still retaining them as ailerons. He produced a new wing that incorporated the Fairey Patent Camber Gear. This Gear comprised hinged flaps that ran along the full trailing edge of both wings. These flaps acted as ailerons but could be wound down by means of a differential control system to increase lift of the wings. This increase in lift enabled a heavier bomb load to be carried and also enabled the machine to achieve lower landing speeds.[4] While this made lateral controls heavy, a control wheel being installed, they were effective right down to the stall.[5] For period W/E 22 July to the W/E 27 October 1916, the *Baby 110 Clerget 8134,* was undergoing tests at the Fairey works at Hamble.[6]

The result of this experimentation was a modified machine. While the fuselage was virtually identical to that of the original Sopwith product the wings were of greater span and chord with rounded tips. The tailplane was redesigned with a rectangular fin and rudder. Larger main floats and a larger-capacity tail float were fitted. The end result of these modifications was virtually a new aeroplane, the Fairey Hamble Baby. The name Hamble came from the Hamble River where Fairey's seaplane base was situated.

The Hamble Baby was to be manufactured by Fairey and by Parnall & Sons at Bristol, Parnall producing more Hamble Babies than Fairey. Parnall Hamble Baby and Baby Converts could be distinguished from Fairy-built aircraft as they retained the Sopwith floats, fins and rudders. The Sopwith type *is more suitable for fighting, while the Hamble Type is more suitable for bombing, having a much lower landing speed and being safe to land with the bombs still in place.*[7]

The Hamble Baby was of typical wood and wire construction with fabric covering. The fuselage was virtually unchanged from the Sopwith original, being moderately streamlined at the engine, quickly tapering into a rectangular cross section. Metal panels covered the forward fuselage. The wings had rounded tips and ailerons on top and bottom, running almost the full length of each wing. The rudder was

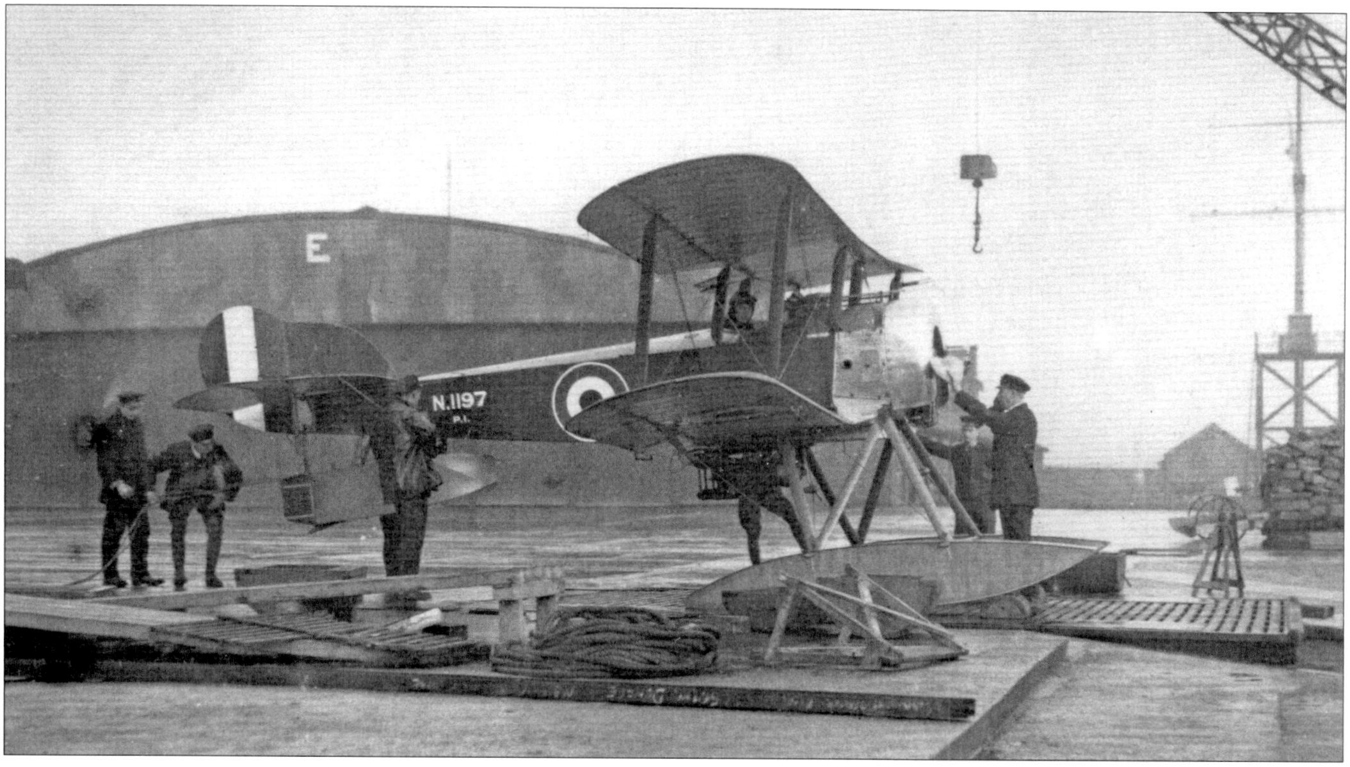

Above: The pilot is in the cockpit of No. N1970 and there is a bomb under the fuselage. The personnel in the photograph may be readying the machine for a patrol. Note the camouflaged hangars in the background.

Below: No. 8134 being readied for a test. The fin and rudder are the original Sopwith variety. The machine has a bomb load suspended under the fuselage at the rear of the rear float cross-bar.

Above: F Warren Merriam, the pilot of this Hamble Baby, shows the length of the hosting sing that was carried in the upper wing centre-section, as his seaplane is lowered from the Jetty at Cattewater in 1917, prior to his test flight to overturn his grounding due to his eyesight. Merriam's autobiography, *First Through the Clouds – The Autobiography of a Box-Kite Pioneer*, was originally published in 1954. It has been republished in 2020. Merriam noted that Cattewater operations consisted almost entirely of anti-submarine patrolling in Short seaplanes. The Sopwith Baby and Fairey Hamble Baby scouts carried a couple of 65-lb bombs, *but were so tricky to land, particularly when there was any sort of sea running, that only Flight Commander Buckland and I used them.*

rhomboidal in shape with the bottom being in line with the bottom of the fuselage. The semi-circular tailplane was mounted on top of the fuselage. The elevators were in two pieces. A large fin of typical Fairey shape completed the tail plane. The two main floats were supported by an "N" system of struts. The tail float was suspended under the tail by four short struts. At the rear of the tail float was a small rectangular water ruder. The Air Intelligence Circulars stated that this *machine resembles very closely the Sopwith Baby Seaplane, which, however, has a rectangular tailplane and fin.*[8]

In early 1916 the policy of using landplanes to conduct over-water patrols was being discussed. Capt A.V. Vyvyan considered that if *this policy is settled it will be necessary to revise the number of Baby seaplanes on order and substitute land machines. Can an estimate of the number of Sop Baby Seaplanes, and Hamble Baby Seaplanes be furnished to the end of 1917 Stating what Stations they should be allocated to, observing that there are a large number of these machines on order which it may be necessary to cancel.*

He made a point of mentioning that it *is not considered that the* (Sopwith) *1½ Strutters are a sufficiently good fighting machine.*[9] Landplanes that had outlived their usefulness at the front were looked at as providing relief for the floatplanes and flying boat patrols. De Havilland D.H.6 trainers were in use by the end of the war for inshore patrols, but the special conditions of the anti-submarine service led to the development of specialised land machines, such as the Blackburn Kangaroo, for this work.

The first Hamble Baby, N1190, was fitted with a 110-hp Cerget rotary and its testing was reported in May 1917. The following performance was recorded: (see table on facing page)

The report concluded with the statement that "*the Machine is suitable for emergency patrol from well sheltered harbour but it is not considered to be a good fighter.*[10]

In late April 1917, the first "Hamble Baby" Seaplane built by Messers Parnell has also been received at Grain Island. This machine is a single seater fighter with variable camber wings of Geddon-Fairey design.[11] In August 1917 it was reported that a land chassis was now being fitter to a Hamble Baby at Grain. Also, new tail floats were being got out to make them *more satisfactory on the water.*[12] In late May N1190 had completed its trials and *been allocated for service.* The same report carried the note that Hamble Baby N1320 *is so far not considered as good as N.1190. She is very tail heavy and the tail fittings are being altered. Trials are being proceeded with.*[13]

The position of the Sopwith Baby and Hamble Baby seaplanes was considered in June 1917. Capt Vivyan reported that there were a large number of Sopwith Baby and Hamble Baby machines on order but from latest reports *they were not up to their work and were generally out-classed by the latest German Machines. In the last Zeppelin raid it was found that one of these machines from Yarmouth was quite unable to reach the Zeppelin;*

Above: A standard Fairey Hamble Baby with bomb load. Note that the lower wing cockades are not painted over the ailerons/flaps.

Hamble Baby Performance May 1917	
Weight in lbs	
Pilot	180
Petrol & oil	190
Synchronised Lewis Gun	45
2 x 65-lb bombs	140
Sub-total	560
Machine light	1,386
Total weight	1,946
Performance	
Speed at 3,500 & 7,900	78 kts
Cruising speed	63 kts
Climb	
to 3,000 ft in	8 min 30
to 7,000 ft in	26 min 30
Consumption in gal/hr	
Petrol	8
Oil	2½

they might possibly be useful still in the Mediterranean, but they were certainly no good for fighting purposes at home.

Capt Vyvyan said that the matter had already been reported by the Operations Division who had considerably reduced their requirements for this type of machine at the various stations. *It was now under consideration to cut the existing orders considerably and to use up the engines thus set free for aeroplanes, preferably the Ship Camel Type fitted with air bags, which could take the place of the Navy seaplanes for over-sea fighting work.* The best course would be for Blackburns to continue with their contract for Sopwith Babies that would be used in the Mediterranean, and for Parnall to go on with their contract for Hamble Babies that would be used on the south-west coastal Stations. It was suggested that Fairey's order be cancelled and an order for Sopwith Camels substituted. As an alternative it was considered that Faireys should be given an order for more Campania seaplanes rather than Camels. This last was adopted.[14]

In November 1917, Cdr Longmore reported to the Progress Committee that *a land chassis had now been fitted to one of these machines by the Experimental Depot, Grain. The trials so far carried out appeared satisfactory and he asked what should now be done with the machine.* In reply Capt Vyvyan stated that if Eastchurch reported favourably on the machine, the question of converting more to land machines would then be considered. *In the meantime it was proposed to store all Hamble*

Above & Below: This Fairey Hamble Baby appears to have become an instructional airframe. Note that the wings have been cut back to the interplane struts. Norman Thompson N.T.2B No. N2562 in the background indicates that the location is the Training Station at Calshot.

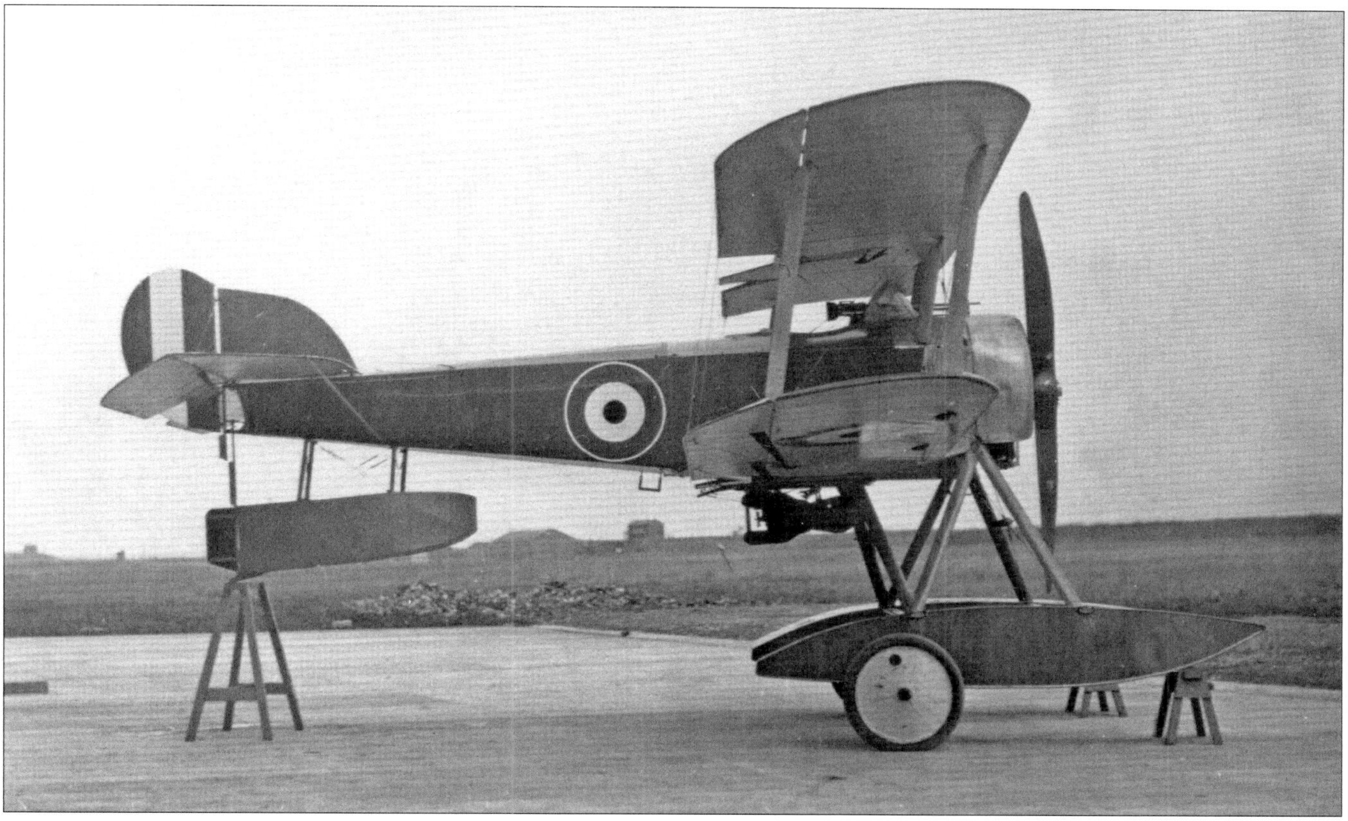

Above: Parnall-built Hamble Baby with the hallmark Sopwith fin and rudder. The machine has a synchronised Lewis gun and a bomb under the fuselage.

Above: Fairey Hamble Baby with the enlarged fin and rudder. Note the machine gun mount in the upper wing centre section.

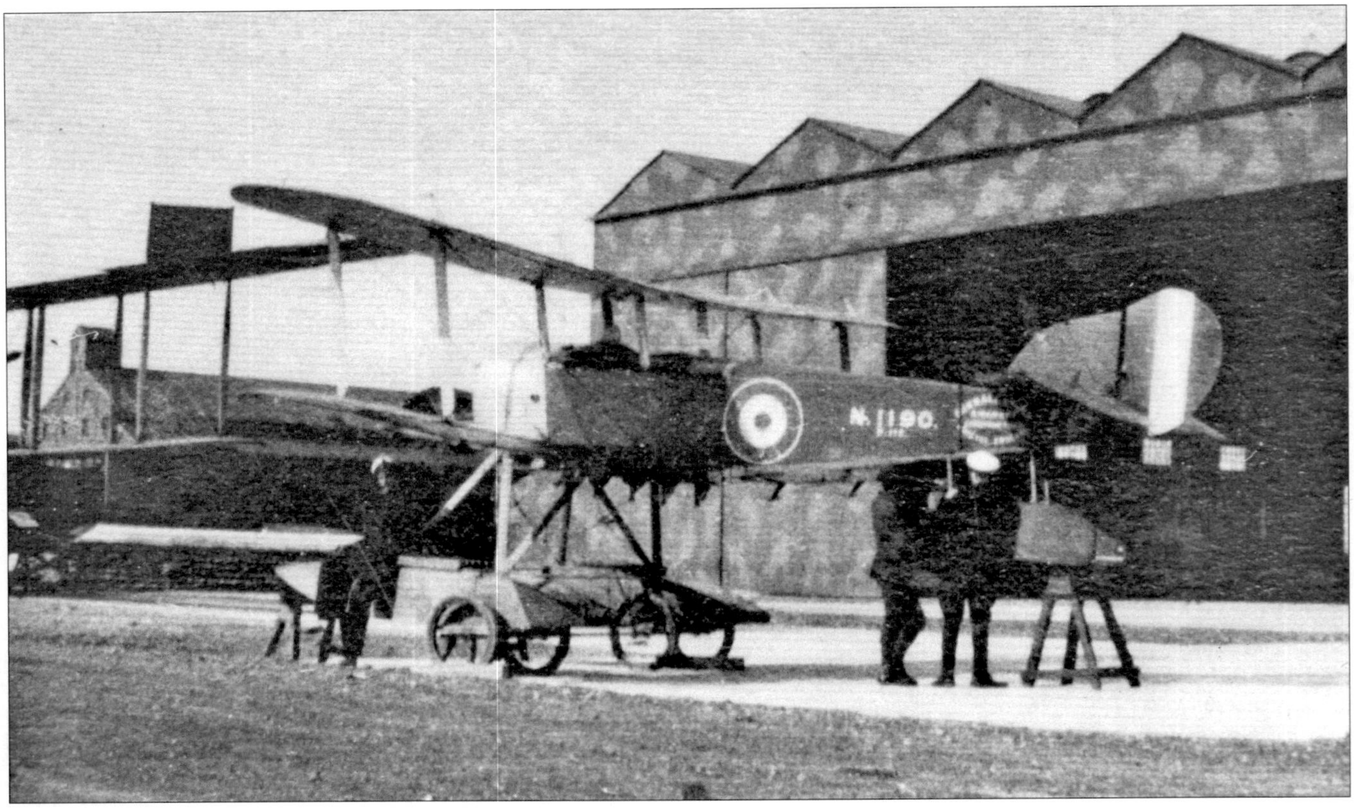

Above & Facing Page, Above: The manufacturer of Hamble Baby No. N1190 is well advertised on the rear fuselage. These photographs were taken at Felixstowe. No. N1190 was first of a batch of 30 floatplanes and was delivered to Grain for testing by rail on 16 Aril 1917. Was at Felixstowe by at least June 1917 and deleted in September 1918.

Baby seaplanes from Parnalls. Those from Fairey's would be allocated to Calshot for training, a proportion going to Otranto as replacements. In reply to a suggestion that the variable camber gear be removed it was decided that this was better retained as it was undesirable to bring such a modification into the production line and the variable incidence was useful for a slow landing, *and might be used in this connection on machines for ship work later on.*[15] The machine under discussion was the Hamble Baby Convert. The Converts were fitted with a very wide track land undercarriage that left the struts in their original position. Skids were fitted together with wheels. All the Hamble Baby Converts were produced by Parnall (N1986 - N2059).

In January 1918, the Committee recorded that reports on the Hamble Baby Converts that had been received were all adverse reports. The remainder of the order for 28 machines was agreed to be cancelled and Parnall to speed up their Avro contract instead.[16] However, the contract was completed; the last 60 Hamble Baby Converts were delivered straight into store at Killingholme.

The Hamble Babies were employed on similar work as the Sopwith Baby. They undertook patrols from coastal based in the UK and the Mediterranean. The seaplane carrier HMS *Empress* operated a number while in the Eastern Mediterranean in late 1917.

Typical examples of the work of the Hamble Baby seaplanes are given below:

23.11.17. N1199. Fishguard. Flt S/Lt De la Rue. Special patrol in search reported submarine. When 12 miles NNW Atrumble Head observed V-wake. The pilot, flying at 1,000 feet, opened up and dived to 500 feet, dropping two 65-lb bombs. The first failed to explode, but the second exploded. Three other machines were sent out from base but nothing further was sighted, visibility being very bad.[17]

27.12.17. Hamble Baby (N)1979 from Hornsea. Anti-submarine patrol of 2 hours 15 minutes to Scarborough. Returned due to heavy rain. Nil sightings. The same day, (N)1469 from the same station, was on an ant-submarine patrol to Filey. It also was terminated due to the rain, the machine only being out for one hour. Nil sightings were the typical answer to the reports of these seaplanes.

24.01.18. N1201. British Aegean Squadron. At 1350 Flt S/Lt Leage sighted a submarine below the surface about20 miles W by S of Cape Tigant. He dropped two 65-lb bombs that fell 10 to 15 feet to starboard amidship. He returned at once, got more bombs and proceeded again to the scene of the action, accompanied by a Short seaplane, but nothing was observed of the submarine.[18]

Above: Fairey Hamble Baby No. N1453, with a synchronised Lewis gun, on the beach at Calshot, 20 September 1917.

Above: Another view of No. N1453, which is surprising, as it was delivered to Calshot in W/E 21 September 1917, and was wrecked and deleted on the 26th the same month.

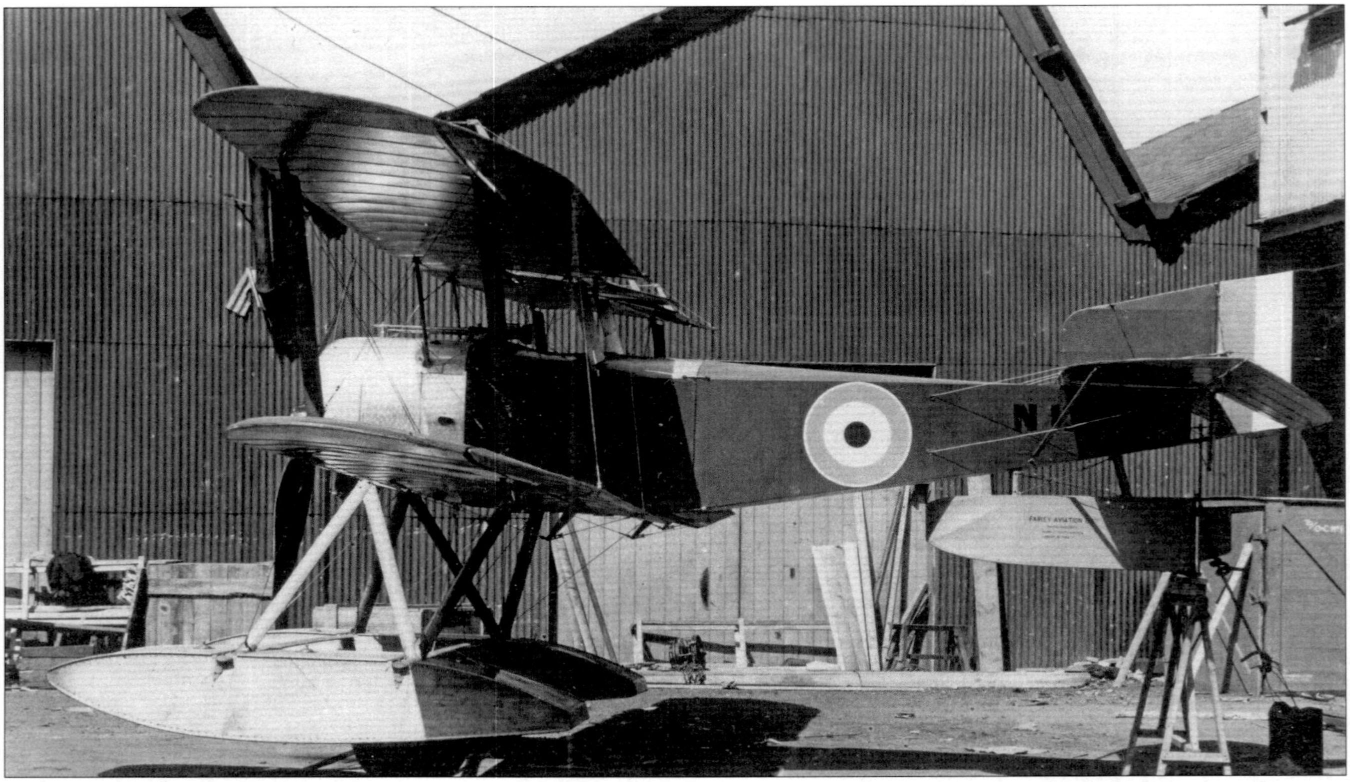

Above & Facing Page, Above: This Hamble Baby is also identified as No. N1453. Note the large hole in the centre-section for the Lewis gun and the hoisting cables.

Above: Launching a Parnall Hamble Baby.

Above: Parnall Hamble Baby and Sopwith Baby at Great Yarmouth. Note D.H.4 A7830, identified by its black and white colour scheme.

19.07.18. N1336. 67 Wing. At 0938 Baby No. (N)1336 was sent to escort convoy approaching Leuca. Machine returned to base at 1038 having escorted convoy to maximum distance possible with this type of machine.[19]

Dunkirk also received Hamble Babies but their activities have not been ascertained to date. For example, Fairey 110-hp Clerget Hamble Babies N1321 to N1322 and Parnall 130-hp Clerget Baby N1194 were allocated to Dunkirk for service in March 1917.[20] In August 1918, Malta had a few *obsolescent single seater seaplanes of the Hamble and Sopwith "Baby" type.*[21]

Action was rare for patrol seaplanes. Most reports end with "Nothing to report." This does not mean that the patrols were without value. These lonely patrols over the unforgiving sea kept the submarines below the surface, hindering their movement and chances to attack shipping. Submarines preferring to be on the surface when attacking if at all possible.

The humble Hamble Baby received one mention in the official history of the RFC, RNAS, RAF – *The War in the Air.* HMS *Empress* had arrived at Port Said in April was operating off the coast of Palestine. On 2 November 1917, during the third battle of Gaza, *Empress'* four Sopwith Baby and two Hamble Baby seaplanes took part in the operations. In the Operational Orders it was stated that there were hostile aircraft based at Ramleh. *If machines are attacked by hostile aircraft when over land, they must make a combined defence. If attacked when over water, they should descend as rapidly as possible, and those machines which are fitted with a moveable Lewis gun through the top plane should then open fire on their opponents.*[22]

Empress had rendezvoused with the *City of Oxford* and the necessary pilots were transferred to the *Empress*. The seaplanes were then hoisted out as follows:

Flt Lt E King in Hamble Baby No. 1209.
Flt Lt Popham in Hamble Baby No. 1210.
Fl Lt H de V Leigh in Sopwith Baby No. 1129
Flt Lt G.D. Smith in Sopwith Baby No. 1038.

No. 1038 immediately began to sink by the tail, slowly turned over and became completely water-logged. Immediate efforts to salvage the machine were given up when the wind vanes on the 65-lb bombs began to unwind, the seaplane being sunk by gunfire from the *Empress*. *This seaplane was fitted with a smaller type of tail float than the other machines, but, at the same time it is considered to have sufficient buoyancy to support this weight.* It was considered inadvisable to send any more machines on the operation as the sea was "confused."[23]

In order to deceive the Turks into thinking that a landing may have been contemplated on the coast behind their right

Above: Close-up of engine and cockpit area of Hamble Baby on Malta, after conversion to a monoplane. Nothing further is known of this machine.

Above: The converted Sopwith Baby No. 8134 with land undercarriage and Fairey wings. Note the Farman trainers in the background. The engine appears to be missing. Photograph probably taken at Eastchurch.

Above: No. 8134 on a snow-covered field. The ailerons/flaps show well in this view. Farman Shorthorn trainers are visible in the left background.

flank, a false landing had been made at Deir el Balah in the afternoon the previous day. The men were reembarked that night when the enemy could not see what was happening. The remaining three floatplanes bombed the railway bridge near Jaljule with six 65-lb bombs but the bridge was not hit. The flying conditions were reported as "very bumpy" with a bad cross wind that made the bombing "Exceedingly difficult." Although the bridge itself was not hit, the line on both sides was hit in several places. On the return of the aircraft the ship sailed to Haifa where they were to attack the oil factory near the port. For this attack the Operational Orders stated that the attack should be *as low as possible (about 500 feet.)*

This time the following machines were hauled out:
Fl Lt Stephens in Sopwith Baby No. 1036.
Flt Lt E King in Hamble Baby No. 1209.
Flt Lt H de V Leigh in Sopwith Baby No. 1129.
Flt Lt Popham in Hamble Baby No. 1210.
Flt Lt G.D. Smith in Sopwith Baby No. 1028.

The first machine had orders to proceed and drop two 65-lb bombs in the vicinity *of the Oil Factory at Hafia, in order to give the Christian employees time to evacuate the factory and take cover.*[24] Popham's engine cut out when he was coming down to attack the factory and he was forced to alight in Hafia Bay where he came under rifle fire. King reported that he saw one machine glide down on to the water about one mile from the head of the Bay. On his return flight he saw the pilot of the disabled machine turning the propeller from the floats. After dropping his bombs Leigh observed the downed seaplane with the pilot standing on the float turning the engine. He circled round again to see if the pilot was successful and saw another seaplane land and commence to taxi towards the ship. He returned to *Empress* but his tail float was damaged in alighting and his machine capsized. No. 1129 was salvaged and hoisted in. Stephens dropping his warning bombs which fell about 50 yards short of the factory. Workers were seen to evacuate the factory. On his return flight he alighted alongside No. 1210. Popham had *No opportunity to destroy my machine owing to being unable to light matches, etc.* Then he swam to Stephen's aircraft while the pair were being subject to rifle fire. Stephens taxied out to sea but was unable to turn into wind to get off the water. In trying to take-off, the extra weight caused the floats of No. 1210 to sink deeper into the water and the airscrew struck the water and broke, as did the longerons and engine bearers. The pilots were rescued by the French destroyer *Coutelas,* but the aircraft was lost, Stephens removed the examination doors of the main floats so as to destroy the machine which sank by the nose. The other seaplane was destroyed by *Coutelas's* gunfire. Four direct hits were reported on the factory *and it is*

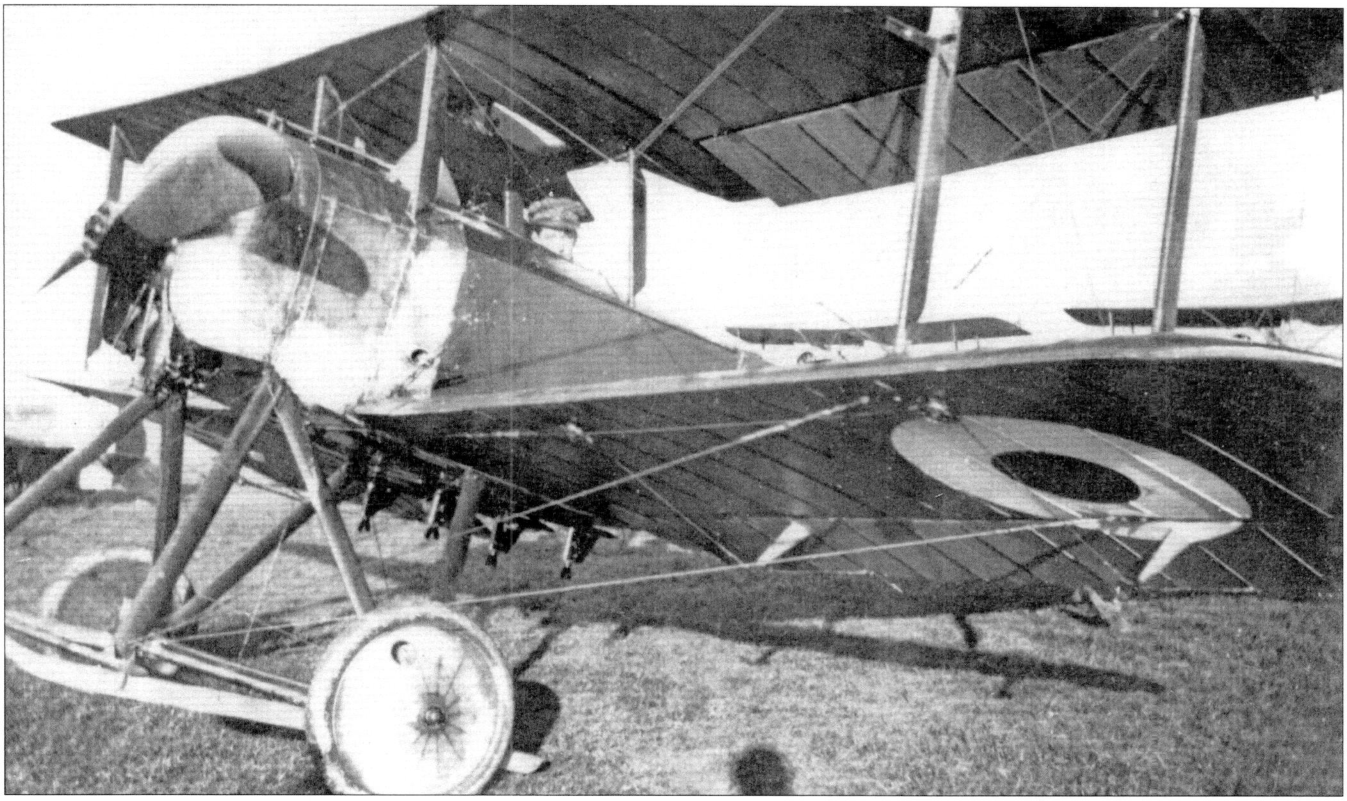

Above: Parnall Hamble Baby Convert. Note the brackets for the bomb rails under the fuselage. The wing cockades now are painted on the ailerons/flaps.

Above: Parnall Hamble Baby Convert No. N1994 was one of the few Baby Converts to see service. Delivered to Yarmouth in March 1918, it served until the War's end. Photo taken 13 June 1918.

Above: This is possibly No. N1994. Note the large oval mounting behind the engine, the purpose of which is unknown.

considered that great damage was caused. Empress then returned to Port Said, it being too late to carry out the third part of the operation.[25]

Fredrick Warren Merriam, always known as Warren, was a pioneer flyer who was an instructor at the Bristol school. He had trained some 200 pupils, of whom about 142 were from the armed services. He entered the RNAS after the outbreak of the war as a civilian instructor due to his poor eyesight. He continued to train aviators, this time for the Navy. After a crash in a Grahame-White biplane, he was sent before a medical board where the specialist who tested his eyes was the one who had turned him down earlier before his commission into the RNAS was granted. Told he had to stop flying he saw a civilian specialist and with the help of Sir Godfrey Paine, then Director of Naval Air Services, he went before another board and unfortunately the result was the same. Transferred to Cattewater he found a friend in the commander, Capt West, who thought of a way of smuggling him back into flying. After being examined by another specialist and the pulling of many strings, he was told that after he made a solo trial flight at Cattewater, his situation would be decided by the Admiralty.

Before I embark on the tale of this trial flight I must explain that I had never flown a sea-plane and that by this time I had been off flying for four months. When I look back, it all seems very amusing; but at the time it was anything but. When the day came I boarded a 130 h.p. single-seater Baby Hamble Scout and was duly lowered on to the water, where I started the engine with the handle inside the cockpit and taxied out. Everything depended on this flight, and, so anxious was I to make a good impression, that I made the cardinal error of omitting to secure the starting-handle. Noticing that here was a large vessel with a tall mast in front of me, I pointed the nose to one side, opened the throttle smoothly and began my take-off. The difference between land and sea-planes in taking off was immediately obvious. The ruffled surface of the water made it more difficult, and the whole feeling was far sloppier. As we became airborne I saw that I had not allowed the ship enough room, and, if I had not gone into an immediate vertical bank, I should have hit the mast with my wing-tip. "Come on Merriam," I muttered, and, then, having recovered from this temporary set-back, I proceeded to climb. Shock number two came when I went to flatten out. The stick would not go forward. A quick look down and there was that blank starting-handle which I had forgotten to secure jamming the forward movement of the stick. There was only one way out – to pull the stick back and risk stalling the machine to release the handle. To do this I had to undo my safety-belt, whip back the stick, and then bend down, disengage the handle and push the stick forward before the Scout had time to stall. With my heart in my mouth, I managed is successfully. Without worrying about my loose belt, I now settled down to some real flying. For the next ten minutes I proceeded to put the machine through her paces, finishing up by cutting out the engine at 2,000 feet and landing just in front of the spot from where I had taken off.

Fairey Hamble Baby Serial Allocation			
Serial	Contract No.	Manufacturer	Notes
N1190-N1219	AS7718/17	Parnall	Most 130-hp Clerget.
N1320-N1329	AS4765/17	Fairey	c/n F.129 – F.138.
N1330-N1339	AS4765/17	Fairey	c/n F.139 - F.148. N1334, N1338, N1338, & N1339 for Greek Government.
N1450-N1479	AS4765/17	Fairey	c/n F.149 - F.178. N1455, N1467, N1468, N1470-N1472 to Greek Government.
N1960-N1985	AS37123	Parnall	N1690, N1961, N1964, N1965, N1978, N1979 for Greek Government.
N1986-N2059	AS37123	Parnall	Fairey Hamble Baby Convert

Everyone thought that he had done well but he had to hide his smile when someone remarked at the "cute" stunts he had performed at the start of the test. Notwithstanding, he was found fit to fly and flew Short and the Baby seaplanes from Cattewater on anti-submarine patrols.[26]

By 31 October 1918, only 18 Hamble Babies were on the RAF charge. Nine were on Home Stations and nine in the Mediterranean.[27] For an aircraft that had been conceived as a Zeppelin fighter, the Hamble Baby ended up in the unglamorous job of patrolling the endless sea, a job it carried out until the end of the war.

Endnotes

1. RNAS Anti-submarine Report No.5, Oct 1917. TNA AIR1/244/223/11/5.
2. TNA AIR1/337/15/226/246.
3. Weekly Report Experimental 23 – 30 June 1916. TNA AIR1/337/15/226/246.
4. Patent No. 132,451 was applied for on 19 May 1916.
5. Fairey continued the work of the Varioplane Company who had been working for some years on a means to make a flexible wing surface. Harold Penrose wrote that although others had been working on the problem, it was Fairey who approached the problem as a practical engineer. (*British Aviation – The Great War and Armistice,* Putnam, UK, 1969). In the post-Armistice downturn in the aviation industry, Fairey offered to sell the variable camber gear to the Government for £10,000. The offer was turned down which turned out lucky for Fairey as his patent continued to provide returns to the company for many more years.
6. TNA AIR1/337/15/226/247.
7. Minutes of 2nd Air Department Weekly Progress Meeting, 22.05.1917. TNA AIR1/270/15/226/113.
8. Intelligence Circular 14. Detailed descriptions of the Chief British and German Machines. TNA WO158/994.
9. A.V. Vyvyan 6.06.17. Fighting Aircraft – Defence Fights S.E. Coast. AIR1/301/15/226/151 Part 2.
10. 14th Weekly Report of the Controller, Technical

Hamble Baby Specifications		
Source:	1	2
Span		27 ft 010 in
Lengtht		23 ft 4 in
Height		10 ft 4 in
Airscrew		A.D.555
Wing Area, ft²	256	256
Weight in lbs		
Empty	1,386	1,386
Fuel & Oil	195	195
Military Load	185	185
Crew	180	180
Gross	1,946	1,946
Speed in kts		
at 2,000 ft	78	78
at 6,000 ft	78	70
at 6,500 ft		
Climb to		
2,000 ft	5.5 min.	5.5 min.
6,000 ft	25 min.	25.0 min.
Ceiling in ft	10,000	7,600
Endurance in hrs	2	2
Engine	110 Clerget	110 Clerget

Source:
1. Air Department Chart: Performance of Seaplanes & Flying Boats. 31.10.1917. Copy in JM Bruce Collection Box 26, RAF Museum. Date of Hamble Baby test report 10.05.17.
2. Chart Performances of British Seaplanes 1917–Dec 31 1918. Sheet 2. Trial Report of May 1917.
The cost of a Hamble Baby was £1,175.00.00.

Above: Another Parnall Hamble Baby Convert at what appears to be a training station. The nose of the machine appears to be painted.

Above: This Fairey Hamble Baby shows how deep the floats were submerged when the engine was under power. This machine has a synchronised Lewis Gun and two bombs as armament.

Above: This photograph is not of good quality but shows an active Fairey Hamble Baby.

Department for the W/E 23.05.1917. Copy in RAF Museum J.M. Bruce Collection Box 25.

11. 10th Weekly Report of the Controller, Technical Dept for the W/E 25.04.1917. TNA AIR1/2094/207/12/10. The reference to "Geddon" has not been able to be determined.
12. Minutes of 14th Meeting of Royal Navy Air Department Progress Committee for 14 August 1917. AIR1/270/15/266/113.
13. Controller, Technical Department Report of 23.05.1917. Copy in RAF Museum J.M. Bruce Collection Box 25.
14. Minutes of 6th Meeting of Royal Navy Air Department Progress Committee for 19 June 1917. AIR1/270/15/266/113. The matter was still being discussed in March 1918.
15. Minutes of 17th Meeting of Royal Navy Air Department Progress Committee for 4 September 1917.
16. Minutes of 33rd Meeting of Royal Navy Air Department Progress Committee for 22 January 1918.
17. RNAS Anti-submarine Report, Nov 1917. TNA AIR1/244/223/11/5.
18. TNA AIR1/271/115/226/120.
19. RAF Mediterranean District Monthly Report of Air Operations July 1918, RAF HQ, Malta. Copy in NARA RG45 Entry 520 Box 138.
20. Allocations made by N.A. on 28 March 1917. TNA AIR1/301/15/226/151 Part 2.
21. "Report of Present Position of Royal Air Force in Mediterranean." August 1918. TNA AIR1/649/17/122/409.
22. Operation Orders by Commander Seaplane Squadron. 31.10.17. TNA AIR1/1708/204/123/74.
23. Report of operations by Commander Seaplane Squadron. 03.11.17. TNA AIR1/1708/204/123/74.
24. RNAS Operations Report No.47. Reprint by Navy & Military Press, UK, 2019.
25. Reports by pilot's and Seaplane Squadron Commander. TNA TNA AIR1/1708/204/123/74.
26. Merriam, F.W. *First through the Clouds,* B.T Batsford Ltd, UK. 1954. P.118.
27. Appendix XLI. *The War in the Air, Appendices.* Reprint by Navy & Military Press, UK.

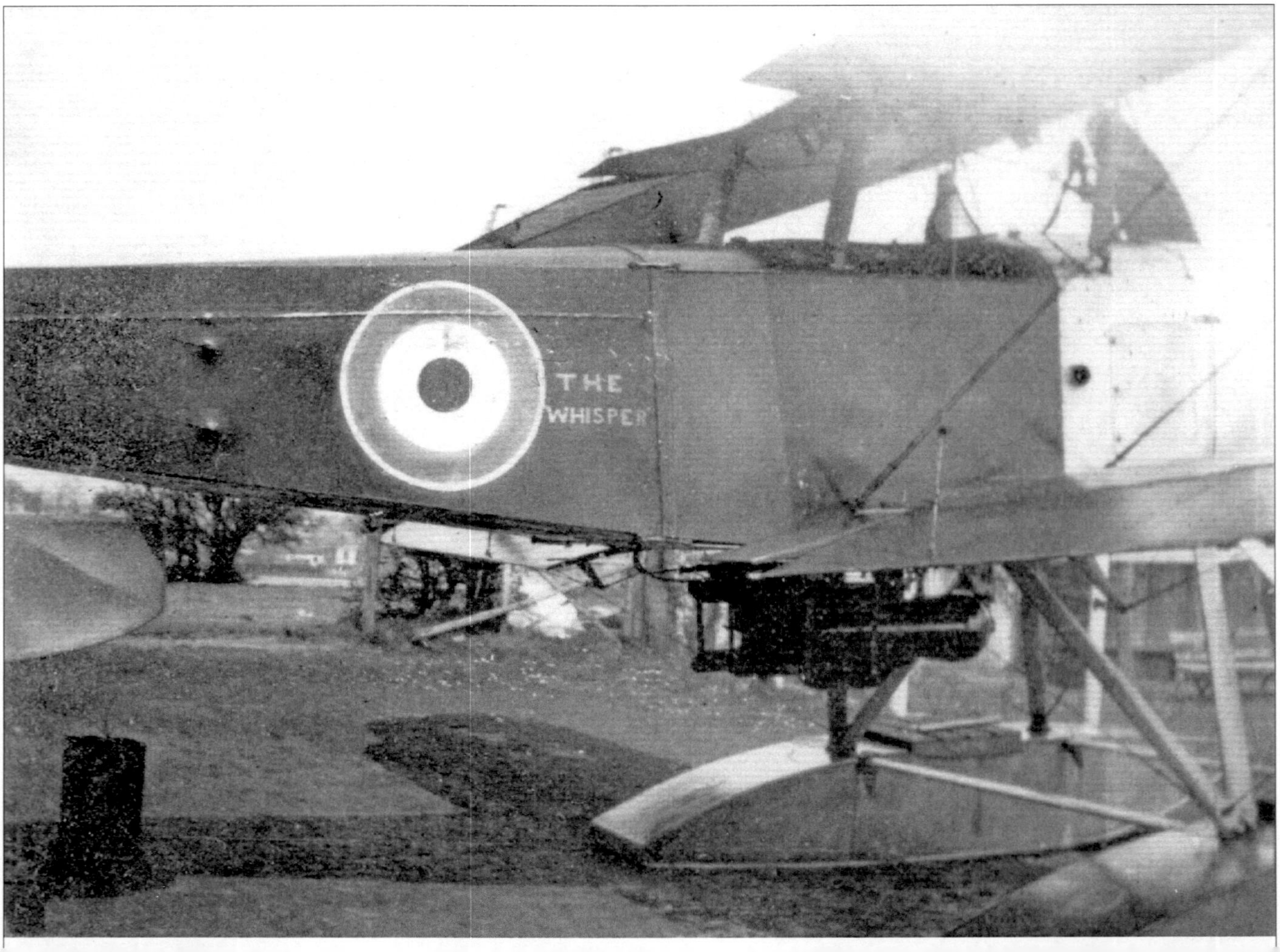
THE
'WHISPER'

Above: Cockpit view of a Hamble Baby.

Facing Page, Above: Another poor quality photograph showing an active Fairey Hamble Baby.

Facing Page, Below: Fairey Hamble Baby *The Whisper*, at Hornsea Mere. Note two bombs under fuselage. Horsea Mere is the largest freshwater lake in Yorkshire and was used as a RNAS base from September 1917. Two slipways and two Bessonneau Hangars were erected and patrols along the coast were flown from here. The HQ was at Killingholme.

Below: N1997 Hamble Baby Convert with a head rest and large windscreen and no armament. This machine served at Cranwell from the W/E 30 November 1917.Note the combined skid and wheel undercarriage.

Above: N2002 carries the stencil "P1/17/" under the serial. No. N2002 was the 17th aircraft of the batch N1986 to N2059. Note the synchronised Lewis gun. This machine was delivered to Cranwell on 27 November 1917 and was still there on 1 April 1918, when the RNAS was amalgamated into the RAF, and the unit became 201/2 TDS Cranwell.

Above: N1190 taxying, with a single bomb and synchronised Lewis gun armament. The Hamble Babies did inshore anti-submarine patrols.

Above: N1190 has small two-wheel dollies attached to its floats. When at rest the machines had to be propped up with wooden horses as shown here. No. N1190 was the first of a batch of 30 Hamble Babies from Parnall. It was at Felixstowe by 3 June 1917, and was deleted on the W/E 12 September 1918. Note the camouflage on the hangars in the background.

Below: A Hamble Baby lifted by a mobile crane. Note the two-wheel dolly attached to the floats.

Above: N1195 appears to have a telescopic sight mounted to starboard in front of the cockpit. This Hamble Baby was sent to Mudros on 10 October 1917. It served on HMS Ark Royal and Peony and was still active in March 1918.

Above: A proud airman poses with his "Baby." This machine is mounted on another type of dolly.

Above: The first Fairey Aviation Co manufactured Hamble Baby No. N1320 of a batch of ten. Delivered to Grain by road on 11 May 1915, it was at Killingholme by the W/E 29 December 1917.It was deleted on 6 April 1918 after a crash on 30 March. Note the engine turned metal panels, and the steel tube cabane struts.

Below: This Hamble Baby on the shore appears to be undergoing maintenance. On the original print the twin Lewis gun barrels may be seen projecting over the top wing centre section.

Above: This Hamble Baby Convert appears to have a large cylindrical metal pipe(?) projecting from the fuselage bottom, immediately behind the engine mounting. This photograph shows the wide wheel track of the Hamble Baby Convert.

N1190 (Facing Page, Top)

Fairey Hamble Baby N1190 was the first of a botch of 30 constructed by Parnall & Sons. This machine has the mounting for another Lewis Gun firing over the airscrew arc. The Lewis gun was so mounted for anti-Zeppelin operations. After testing at the Isle of Grain from April to June1917, it appears that N1190 spent all of its service at Felixstowe. Deleted September 1918.

N2002

A Fairey Hamble Baby Convert with land undercarriage, N2002 was one of a batch of 70 contracted to Parnall and Sons. Delivered to Cranwell in the W/E 7 December 1917, it survived until at least 1 April 1918. The majority of this batch were delivered to storage at Killingholme.

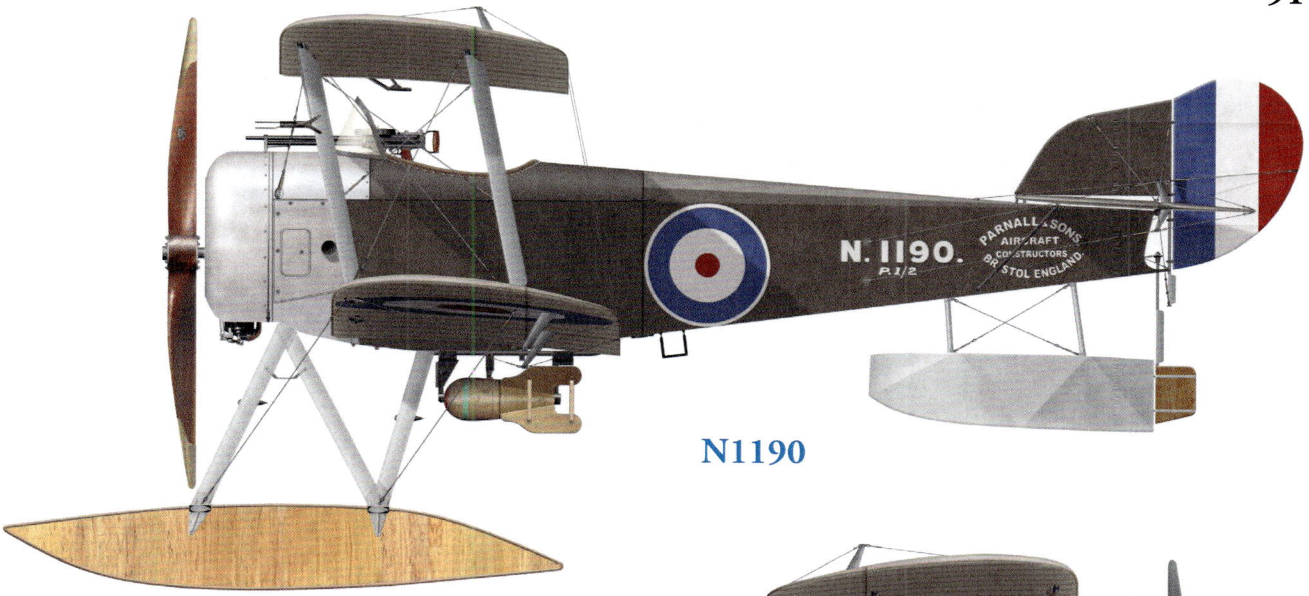

N1190

RONNY BAR PROFILES
ronnybarprofiles@gmail.com

N1994
Delivered to Killingholme store in November 1917, N1994 was at Yarmouth by March 1918. Became 490 Flight. Deleted in June 1918.

N1453
N1453 was a Fairey constructed Hamble Baby of a batch of 30. Delivered to Calshot W/E 21 September 1917, it did not survive long, being wrecked and written off on 26 September.

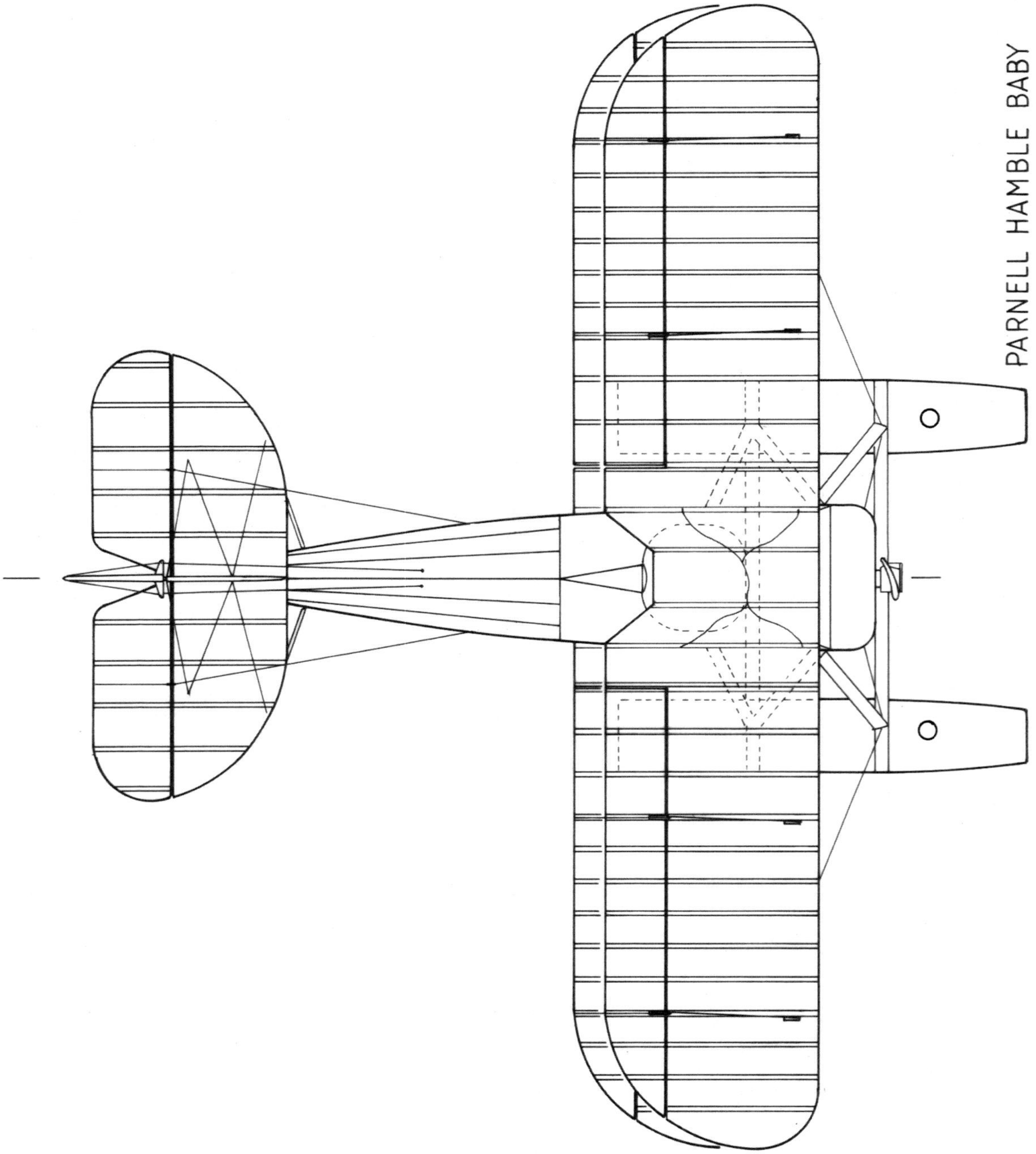
PARNELL HAMBLE BABY

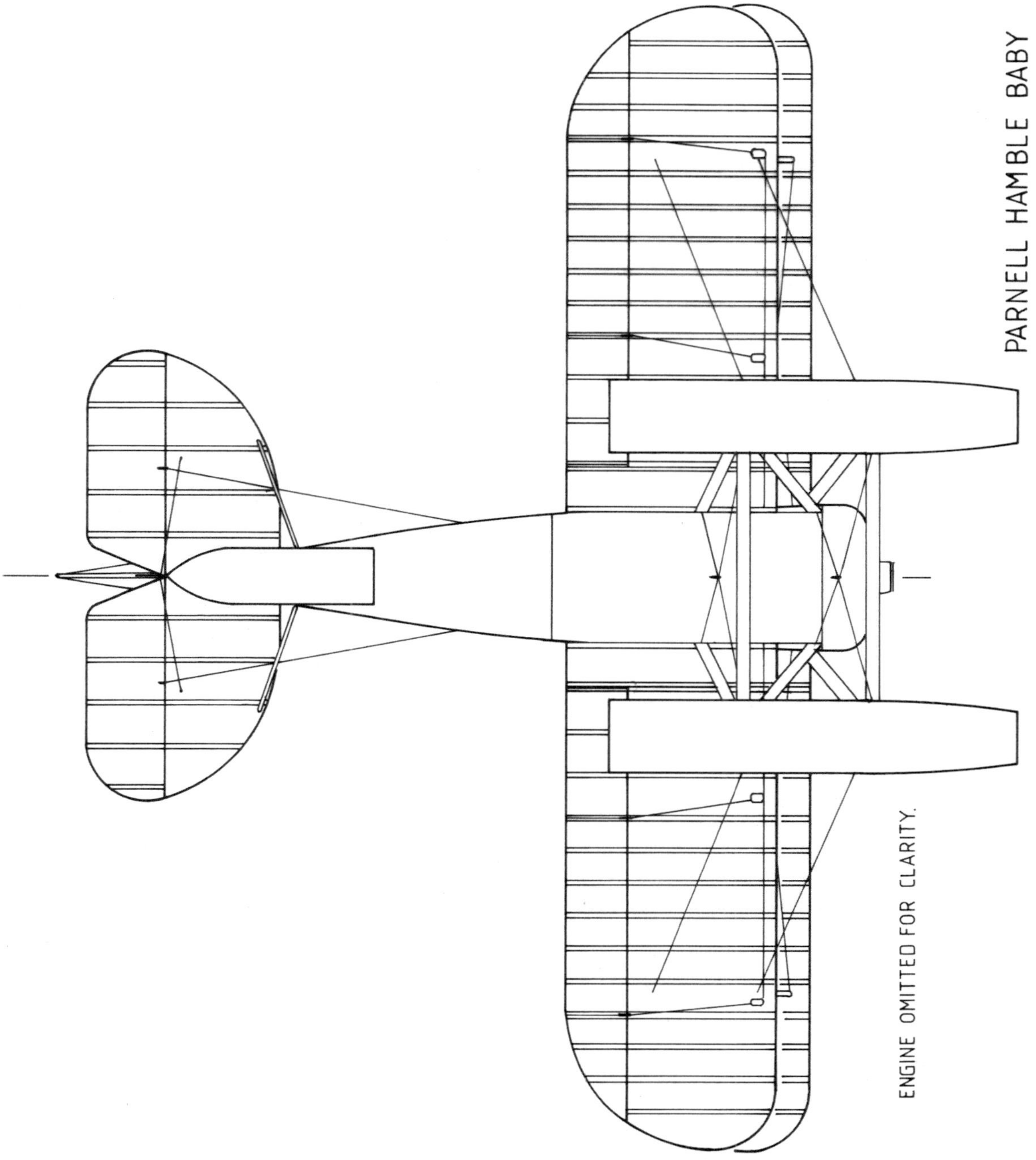
PARNELL HAMBLE BABY
ENGINE OMITTED FOR CLARITY.

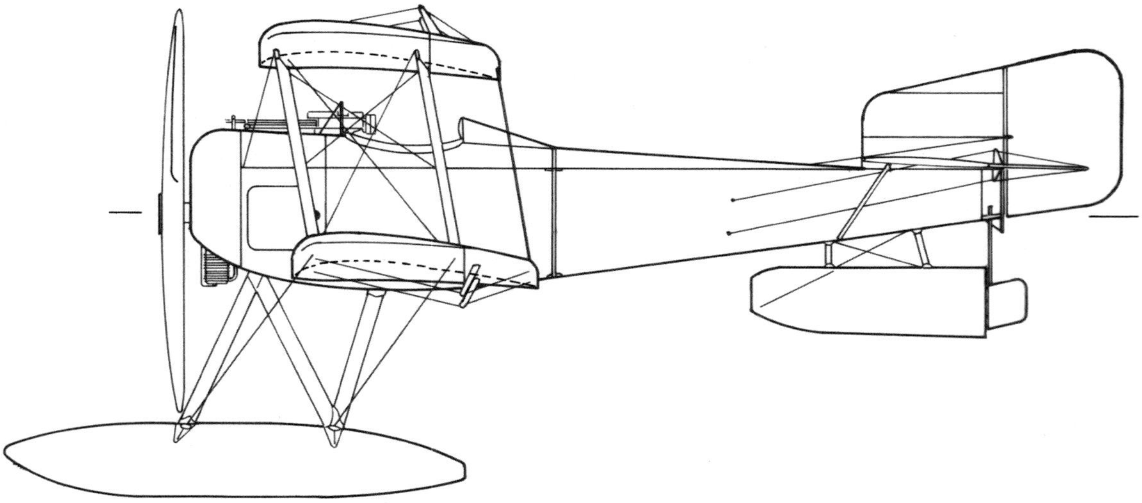

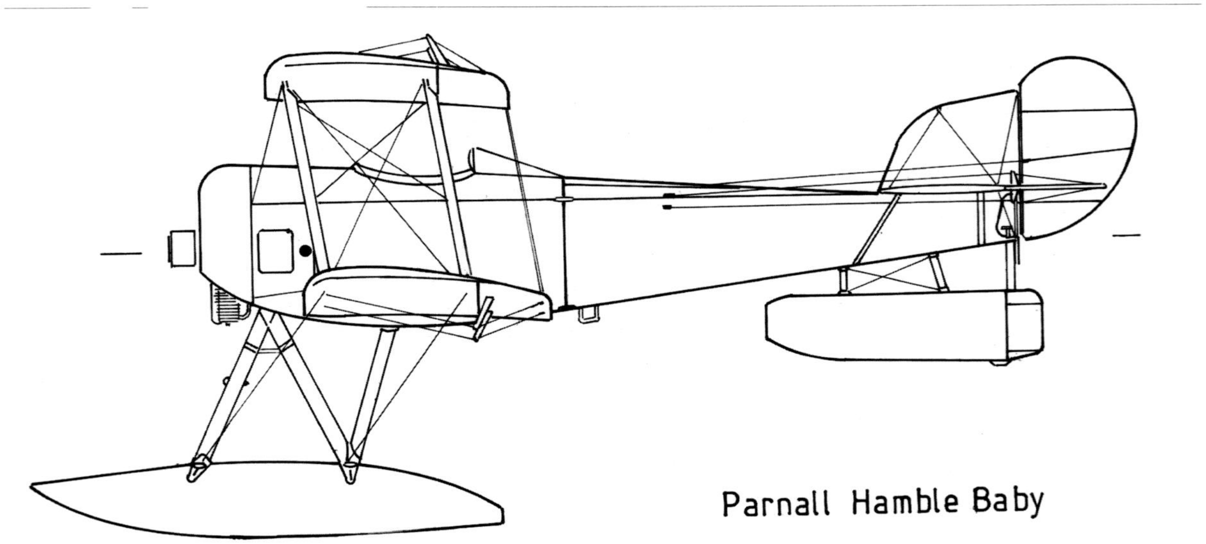

Parnall Hamble Baby

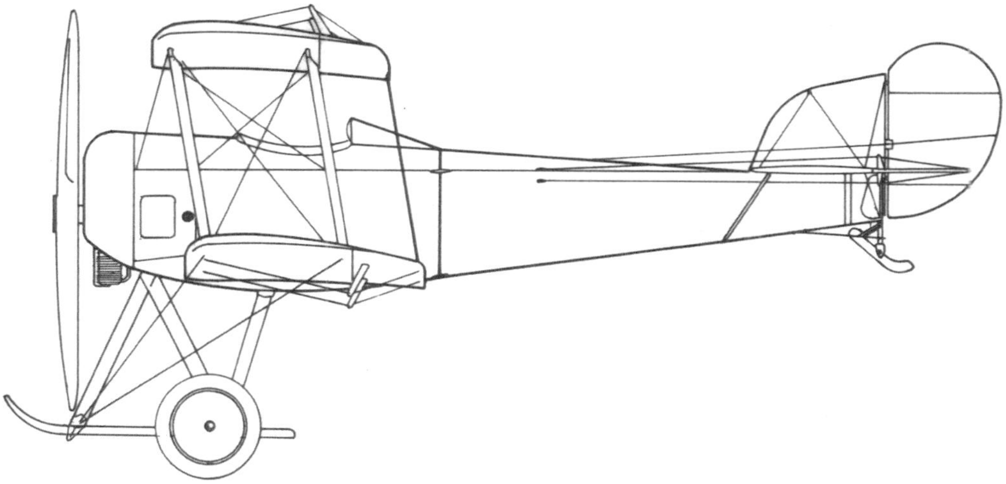

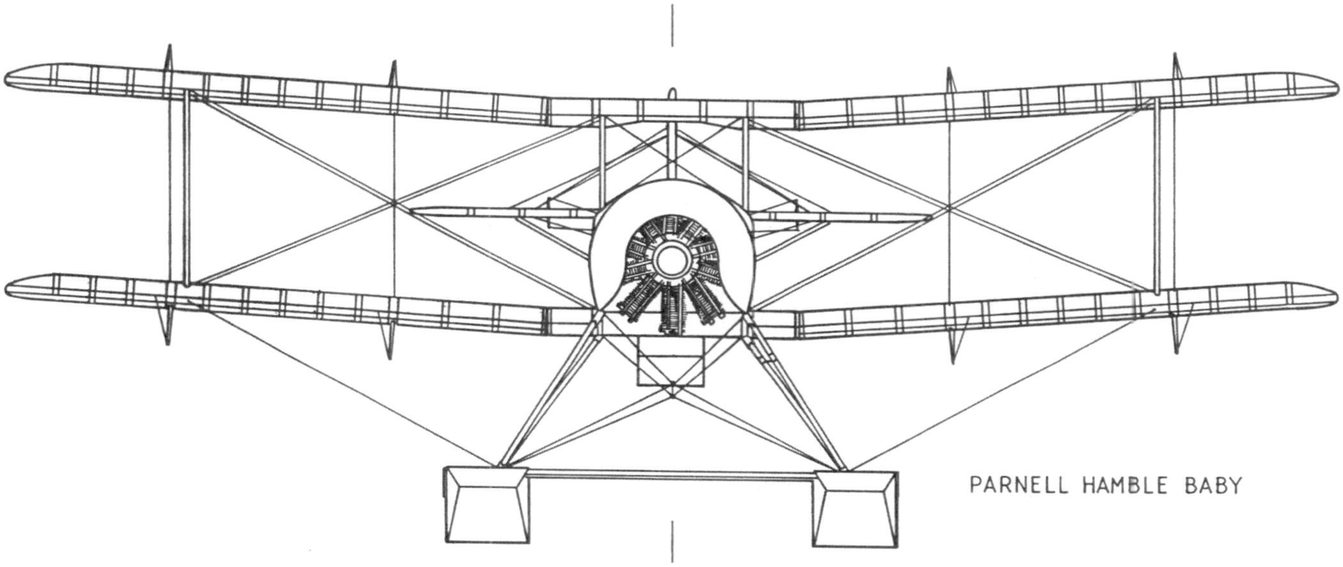

PARNELL HAMBLE BABY

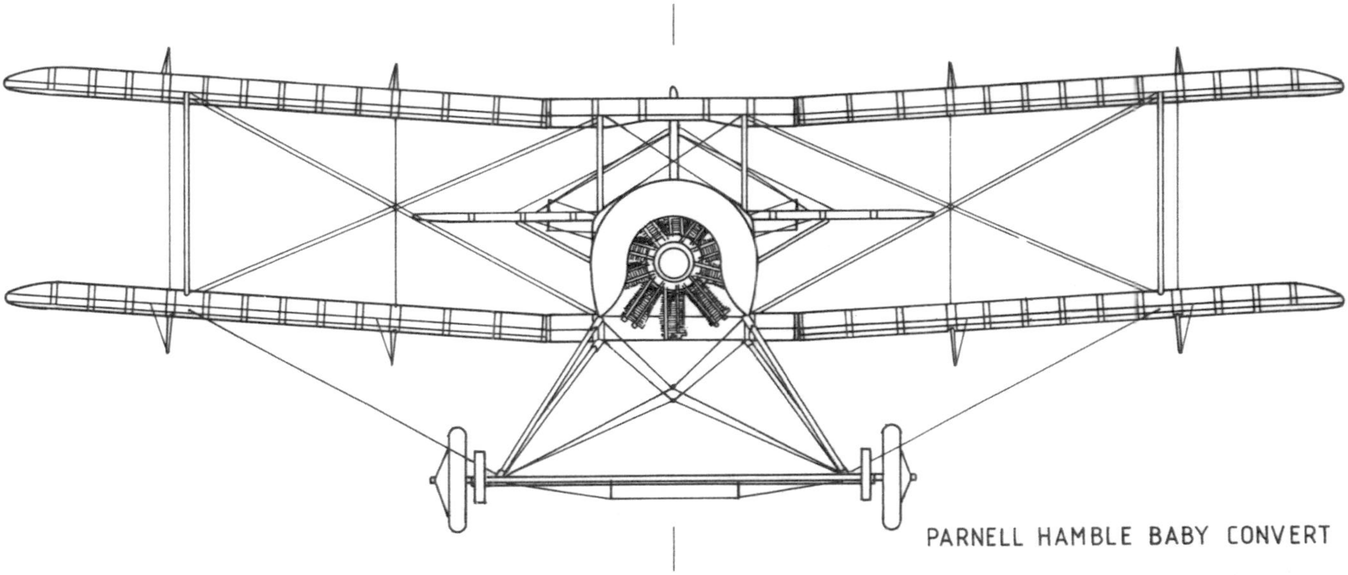

PARNELL HAMBLE BABY CONVERT

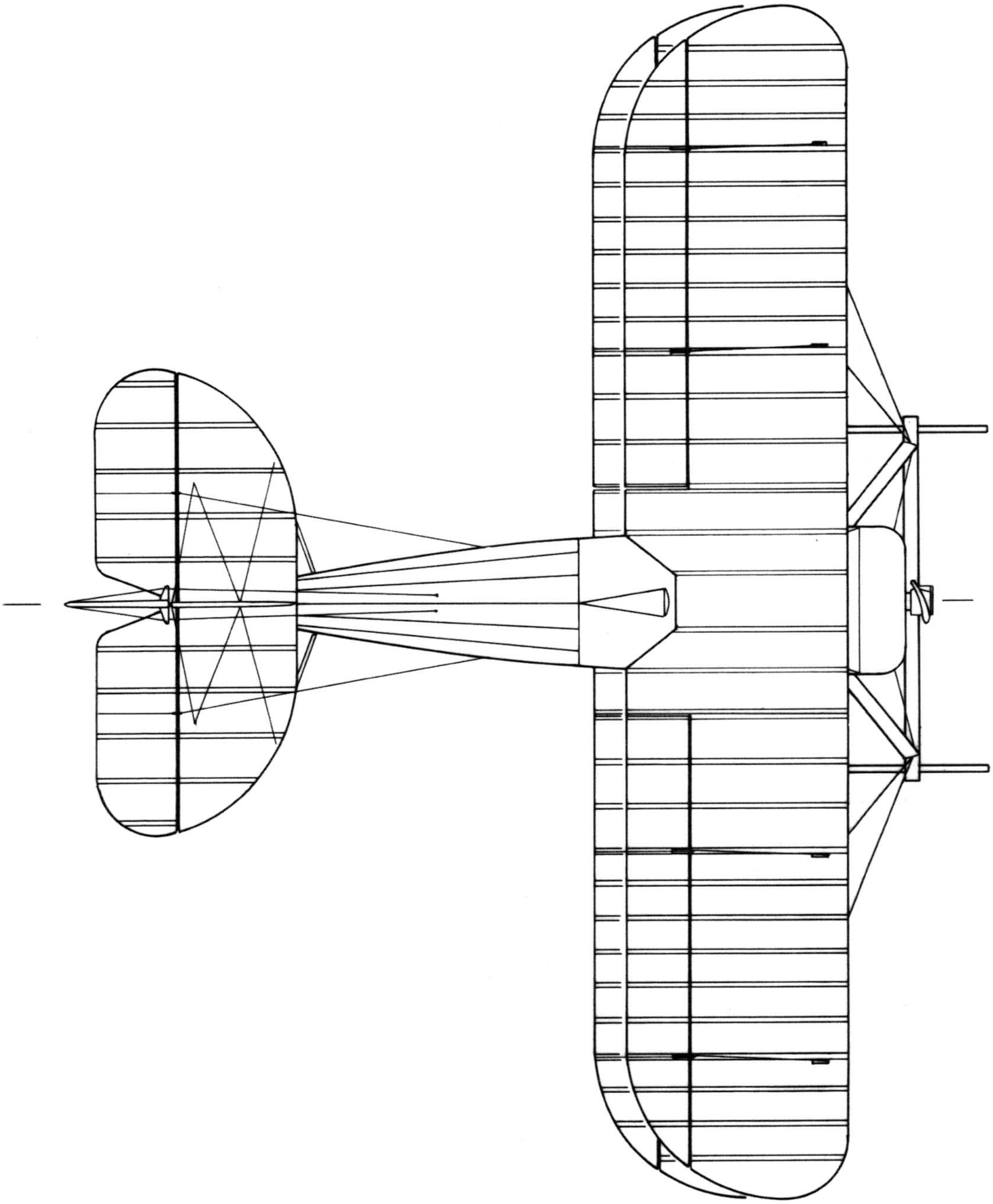

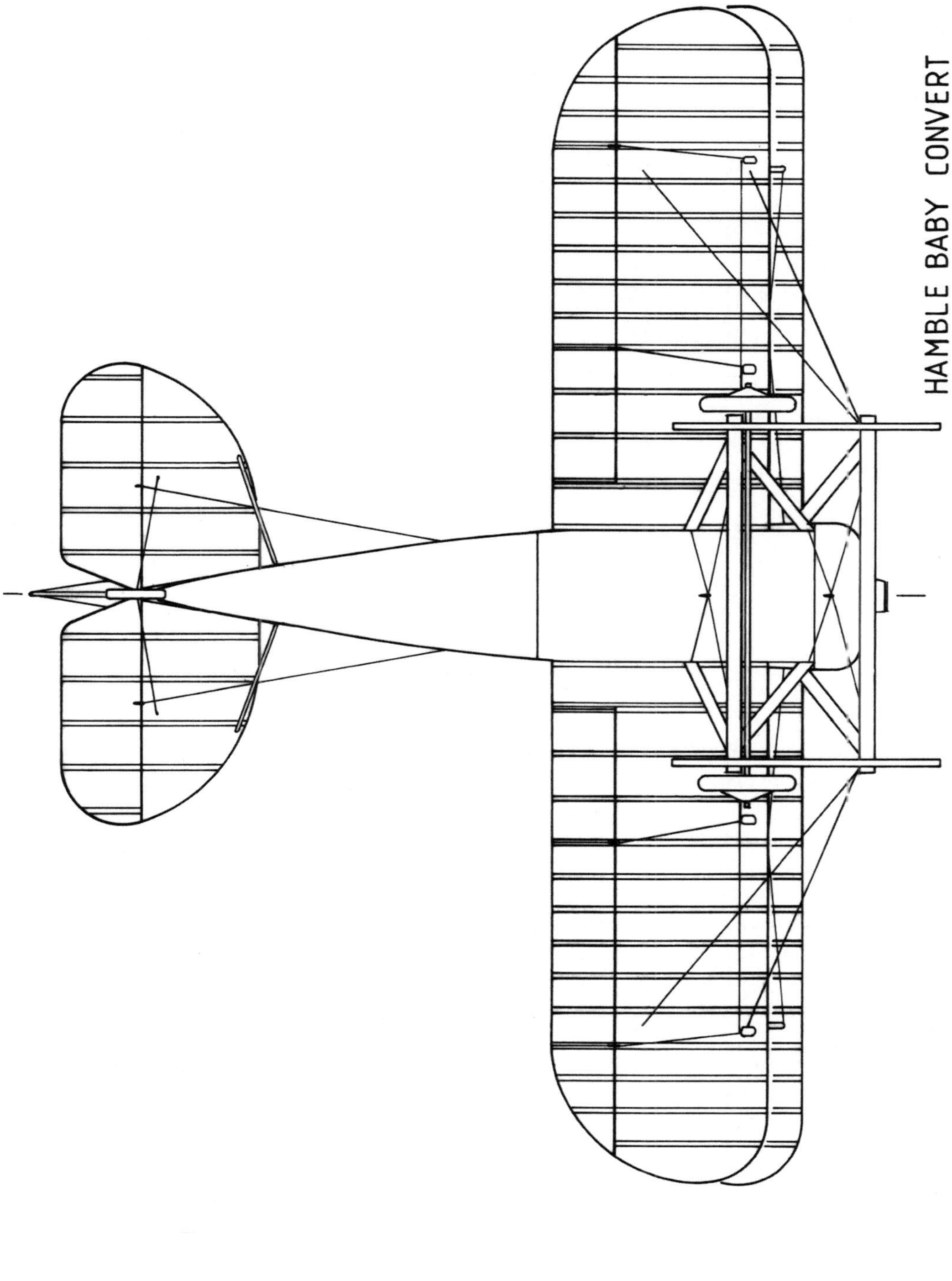
HAMBLE BABY CONVERT

Chapter 5: The Fairey F.127 / N9

Above: Starboard view of the F.127/N9 as first rolled out. Fairey patent camber changing gear flaps were on the full span of the lower wings only

The Campania was followed by the F.127, built in 1917, it was the first in the line that was to lead to the successful F.III series. A chart of "Experimental Seaplanes on Order" dated 9 May 1917, recorded that the *2 Seater Tractor 190 R.R. type 3 (old)* was considered for the *New type 2a* and that good results were expected.[1]

The Fairey F.127 was allocated the serial N9, both these designations being used in official documents. The N9 was a small, neat patrol floatplane capable of being launched from seaplane carriers using the trolley-separation system developed for the *Campania*. Its wings could be folded for stowage aboard ships and full use was made of the Fairey Patent Camber Gear with trailing edge flaps fitted to the upper wing between the ailerons and to the entire trailing edge of the lower wing. Radiators were attached each side of the nose and the exhaust pipes of the Rolls Royce Falcon I engine passed through the centre-section.

Although the type did not result in a production order it was to serve at Hamble and the Isle of Grain. The 20th Weekly Report of the Controller, Technical Department for the W/E 25 April 1917,[2] noted that the first of Fairey's two "Type 3" seaplanes had been received at Grain for test. The machine had a 190-hp Rolls Royce engine and was intended for fleet reconnaissance from seaplane carriers. N9 had arrived on the 20th and was still at Grain in the W/E 5 December 1918, for catapult trials, and it was recorded as a "Fairey Type 3A."

N.9 undertook the catapult experiments with HMS *Slinger*, a converted mud-carrying vessel that had been purchased from Lobnitz and Co Ltd of Renfrew, Scotland, before it was completed, and fitted it out with an Armstrong-Whitworth 60 feet-long compressed air catapult. It had been reported in September 1917 that *the first trials of the Armstrong catapult in the barge "Slinger"* had taken place at Armstrong's works on the Tyne. *A mock-up of a seaplane was used for these experiments, which had neither wings nor engine fitted, but in which weights had been arranged to be the same as those of a seaplane ready for use. The trials were quite satisfactory and the mock-up was catapulted at a rate of 30 miles an hour.* The following week a complete machine without engine was to be launched. If the trials were successful then the barge would be sent to Grain for proper flying experiments.[3]

Successful launches were conducted by Lt Col Harry Busteed with the ship at rest and underway. The machine had been specially strengthened to take the shock of catapulting

Above: F Duncanson completed the preliminary designs for the Admiralty's Air Board N.2(a) specification in 1917. The N9 had elegant lines marred by the side radiators.

Above: The long extensions of the upper wing were braced from the king posts situated above the interplane struts.

with additional struts fitted to the forward horizontal tie-rods between the floats. Busteed could not recall the *exact date of the return to the Isle of Grain of H.M.S. "Slinger" with the Armstrong Catapult installed...it was not until about the middle of June, 1918, that complete trials were commenced. The aircraft used was a Fairey (type N.9) strengthened for the purpose. The first trial was carried out at anchor, with the ship head to wind and a launching pressure on the catapult for a velocity of about 35 m.p.h. The aircraft was slightly damaged. Two other launchings were made shortly after the aircraft was repaired; one again at anchor with the ship about 10° off wind. The 3rd. was carried out with the ship under weigh.*

Above & Below: Two more photos showing that the long extensions of the upper wing were braced from the king posts situated above the interplane struts.

After the Armistice the "Slinger" was paid off and later sold; the catapult went with her.

I think I am right in saying there were only three launchings, interest in catapults at that time was not very keen, as the apparatus was considered too big for installation in a ship of war and the fleet's requirements in this direction was covered by launching platforms on the gun turrets.[4]

F.127/N9 was purchased back by Fairey and an official application made "re Fairey Modified Type 3". Fairey applied for a Certificate of Airworthiness (CofA) for F.127modified for carrying passengers or goods. The application was marked "Doubtful." It was thought that there was *likely to be considerable trouble with this machine.* The writer wanted to know if the modifications that had been carried out on the machine would be approved. It was understood that Fairey did not want to use the aircraft for passenger hire and confirmation of this was asked for. It was reported back that F.127 had been seen by officers of D Section and the *actual machine is a combination of the original N9 fuselage, engine and chassis with N10 floats and 3a wings. The gross weight of the N9 on type trials was 3812 lbs. and it is considered desirable that this weight should not be exceeded under the certificate conditions.* Fairey still wanted the CofA. In July it was reported that Fairey no longer wanted a CofA and *no further section action in the matter was necessary.*[5] N9 was rebuilt and re-engined with a 250-hp Maori II and registered as K-103 in May 1919. As G-EAAJ it was sold to the Norwegian Navy in May 1920. By 1927 it had the Norwegian civilian registration N-20. It was finally scrapped in February 1929, after crashing on the River Vorma.

End Notes

1. TNA AIR1/6A/4/47.
2. TNA AIR1/2094/207/12/10.
3. Minutes of the 20th Air Department Weekly Progress Meeting of 25.09.1917. Report of Sqn Cmdr Miley. TNA AIR1/270/15/226/113.
4. "Notes on work carried out at R.N. Station Isle of Grain in connection with Catapults." Gp Capt HR Busteed, 12.03.1931. These notes were sent by Busteed to the Historical Section, RAF, presumably in reply to a request from them.

   In his covering letter Busteed noted that *there is roughly a period of ten years between rediscoveries, where we begin to tramp over the same old country again. It is a little disappointing and really confirms that progress in aircraft is very meagre and would be practically negligible if engine design had not progressed.*
   Flotation gear for aircraft with the fleet is now in the course of resurrection. This was a burning question a little more than ten years ago. TNA AIR1/728/170/3.

Above: With the wings folded back, the N9 shows the two exhaust pipes for the 190-hp Rolls-Royce Falcon I engine located through the centre-section, a recognition feature of the machine.

5. Fairey had applied on 2nd May. TNA AIR2/970.

H.M.S. *Slinger's* Catapult

Harry Busteed recalled that three types were constructed with Compressed Air, Hydraulics and Electricity as a motive force.

The Compressed air apparatus was constructed by Armstrong's at Newcastle on Tyne, was installed in a Steam Hopper, which was taken over by the Admiralty for the purpose.

General Arrangement:

Length of launching rail 60 ft. built up in the form of a big box girder; the rams and pulleys being accommodated inside the girder. The trolley which supported the aircraft was impelled along the rails by means of wire ropes; an arresting device to deaccelerate the trolley at the end of its run being provided. It was possible to obtain a velocity of approximately 60 miles per hour using a very high pressure, but no greater velocity than about 35–40 m.p.h. was ever used with a properly manned aeroplane.

Dummy trials were carried out with an old seaplane loaded to 5,000 lbs. This aircraft with personnel aboard was shot off successfully in the Tyne about the end of September, 1917.

The ship was then sent down from the Tyne to the Isle of Grain for further trials.

The Hydraulic apparatus was never used and was broken up before leaving the works somewhere in Glasgow. I cannot remember the name of the firm.

The Electrical apparatus was, I believe, installed at Hendon and a machine launched from it.

Source: "Notes on work carried out at R.N. Station Isle of Grain in connection with Catapults." Gp Capt HR Busteed, 12.03.1931. TNA AIR1/728/170/3.

Above & Below: Two more photos showing with the wings folded back, the N9 shows the two exhaust pipes for the 190-hp Rolls-Royce Falcon I engine located through the centre-section, a recognition feature of the machine.

Above: The N-9 at Grain.

Above: Launching the N9. With its overhang on each side, the machine was of sesquiplane layout.

Above, Below, & Facing Page: The N9 on the catapult of HMS *Slinger* on 14 May 1918. The N9 was the first catapult launch of an aircraft at sea by the UK. The USN had catapults operating on the armoured cruisers USS *North Carolina* and *Seattle* (1916) and *Huntington* and *Montana* (1917), although the *Montana's* was removed before she carried any aircraft. On the entry of the US into the conflict they were removed to allow the vessels to act in their traditional roles.

Above: The N9 on the catapult of HMS *Slinger* on 14 May 1918.

Below & Facing Page, Above: Launching the N9 from HMAS *Slinger* on 14 May 1918. Although it never carried bombs, the N9 was ballasted for two 112-lb bombs when it was used in the catapult trials with HMS *Slinger*.

Above: Pickle's Fairey at Westgate, 1919. Fairey purchased N9 back and rebuilt it with equal span wings and installed a 250-hp Maori engine. It is seen here at Westgate-on-Sea when it was employed on the *Evening News* special newspaper service. The machine started at the Thames at Blackfriars and took copies of the newspaper to Thanet.

Above: With the sides of the fuselage emblazoned with the Fairey Company name more prominent than the *Evening News* logo, the N-9 is subject to public scrutiny while on the beach.

Above: N9 was to enter the civil register as K-103 and then G-EAAJ. Later it was sold to Norway.

Specifications F.127/N9		
Source	J.M. Bruce data	Jane's 1919
Span upper	50 ft	50 ft
Span lower		
Length	35 ft 6 in	35 ft 6 in
Height	13 ft	13 ft
Chord	5 ft 6 in	5 ft 6 in
Gap	5 ft 7 in	5 ft 7 in
Span of tail	13 ft	13 ft
Areas in sq ft		
Wings	420	420
Ailerons	18 each	18 each
Tailplane	34.2	34.2
Elevators	34.2	34.2
Fin	8.9	8.9
Rudder	9.8	9.8
Weight in lbs		
Empty	2,699	
Military load	216	
Crew	360	
Fuel & oil	537	
Total	3,812	3,812
Fuel	70 gall	70 gall
Performance		
Sea level	90 mph	78 kts
At 10,000 ft	86 mph	74 kts
Landing		38 mph
Climb to		
2,000 ft	4 min 10 sec	
5,000 ft	9 min 20 sec	9.30 min
6,500 ft	18 min 30 sec	
10,000 ft	38 min	38 min
Service ceiling	8,600 ft	
Endurance	5¼ hrs	5¼ hrs

Fairey N9

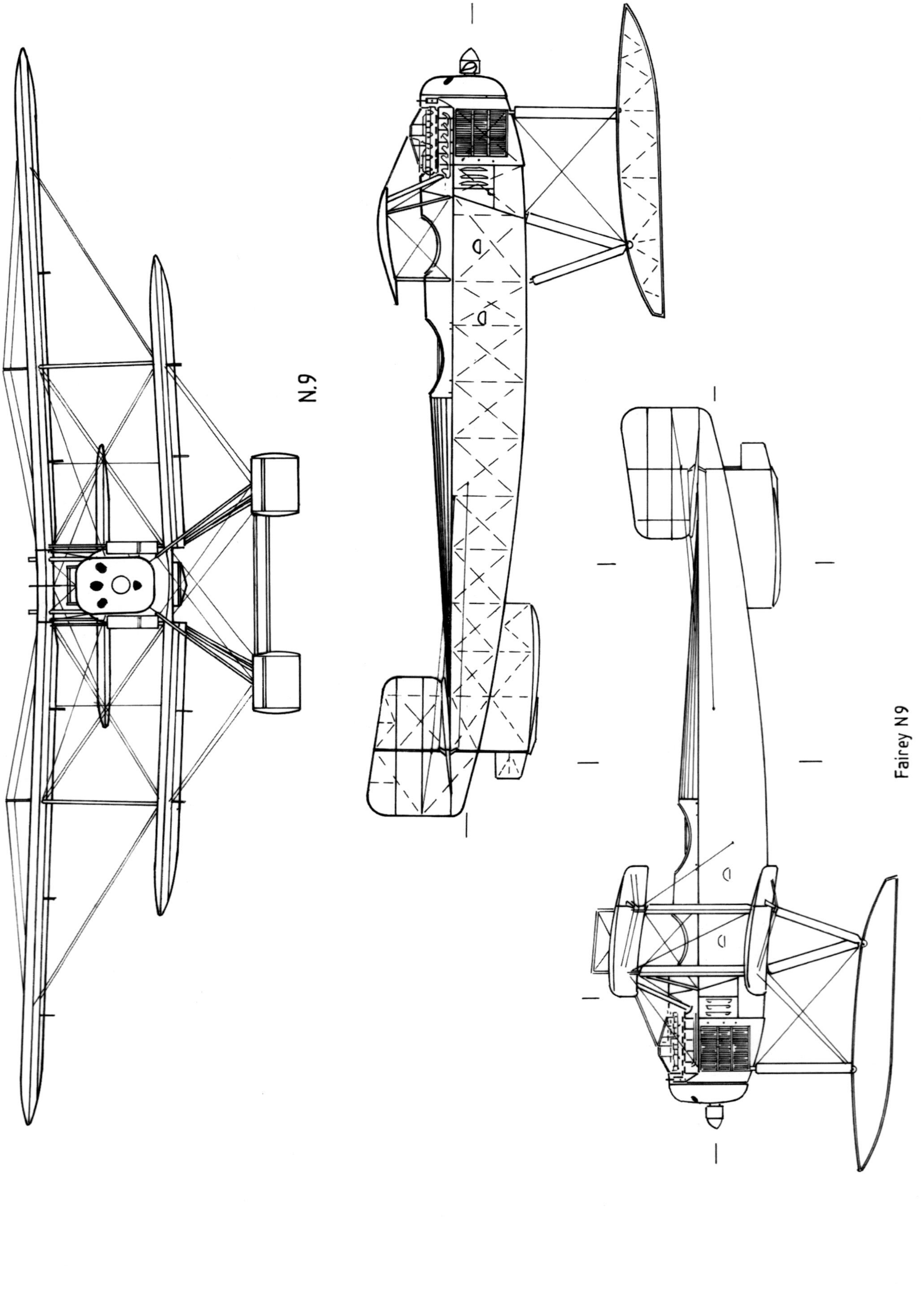

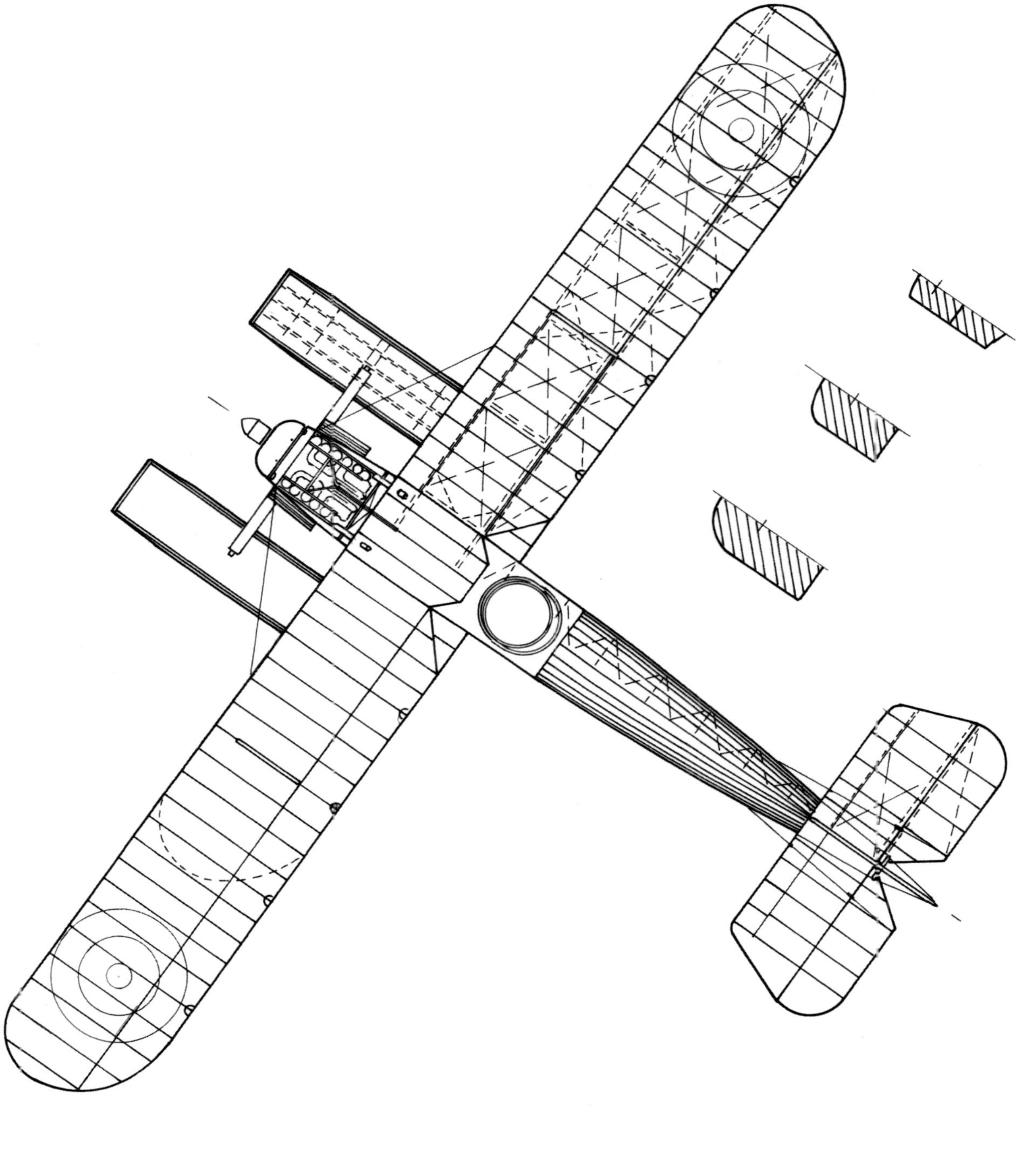

Chapter 6: The Fairey N10 and Type III to IIIC

Above & Facing Page: The N10 at Hayes in its original two-bay configuration as first flown. Built at the same time as the N9 and to the same specification, the F.128/N10 had the 260-hp Sunbeam Maori II engine. Side radiators were used as in the N9. Note the Fairey logo with the c/n F128 in the centre. The white stencil states that the "MAX. USEFUL LOAD 800 LBS." 24 September 1917. Probably taken at the Isle of Grain.

Post-WWI the Fairey Aviation Co advertised its Type III Series as *supplied in quantity to the British Government during the War and* *since its termination. In this series of Machines the Fuselage, Centre Plane, Tail Unit, etc., are standard and adapted to take different wings, chassis and engines of other Machines of the series to suit the particular purpose for which the machine is required. All machines of this type have folding wings fitted with our Patent 'Variable Camber Gear giving high performance and lifting power combined with low landing speed.*[1] Sir Richard Fairey had learnt to promote his products and was to continue to develop the Type III line up to the IIIF ensuring the survival of his company in the inter-war years. The obsolete IIIF was still in service when Japan entered World War II.

The Campania established Fairey as the premier builder of floatplanes for the RNAS, taking that mantle away from Short Brothers. It was not surprising then that they worked closely with the Admiralty Air Department to produce the next generation of seaplane to replace the venerable Short 184, although, as will be related, this was not to take place in the way as first envisaged.

Type F.128/N10

The Fairey N10 had a long, interesting, and complicated life in service and afterwards as a civil machine. It took place in experiments that led to the Fairey IIIA landplane, and with IIIB wings, it entered trials for the Admiralty specification Type N.2(a) and N.2(b) seaplanes.

Constructed at the same time as the F.127/N9 was the F.128 or N10 (so called after its serial number as had been the case with the N9). The N10 had a more powerful 12-cylinder Vee liquid-cooled 260-hp Sunbeam Maori II engine, No.2436, as it was designed as a two-seat patrol seaplane. The biplane had two-bay wings of equal span and they were arranged to fold for stowage. Fairey Patent Camber Gear was fitted. The fuselage was identical to that of the F.127/N9 with the same tailplane and rudder but a larger fin.

A 47-gallon pressure fuel tank was situated under the pilot's seat with a 4-gallon oil gravity tank in the upper plane centre section. A compressed air starter was installed. The average time to get away was 10 minutes. The view from the pilot's seat was not good, otherwise the machine was considered suitable

N 10

Above: Another photo of the N10 at Hayes in its original two-bay configuration as first flown.

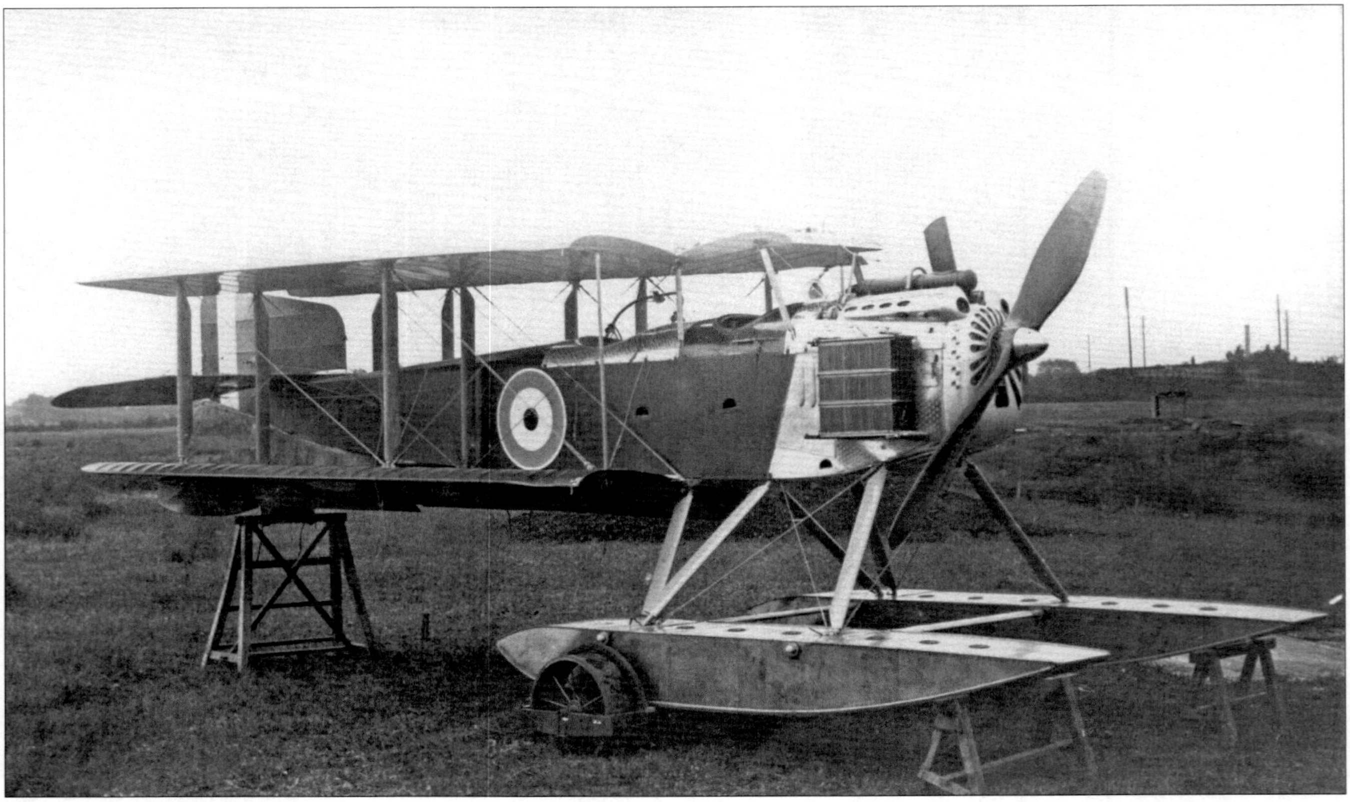

Above & Facing Page: The Fairey patent camber changing fear flaps were folded up on the top wings and down on the lower wings to allow for folding of the wings. Grain photographs 24 September 1917.

This Page: Landplane conversion of the N10 with frontal radiator. This machine became the prototype for the Fairey IIIA shipborne two-seater to replace the obsolete Sopwith 1½ Strutters in RNAS service.

Above: N10 landplane conversion radiator with adjustable radiator screen closed.

Above: N10 landplane conversion radiator with adjustable radiator screen open.

This Page: The N10 was reconverted back to a seaplane with Fairey IIIB type wings and tailplane. Photographed at Grain on 6 February 1918.

Above: The modified N.10 on the water.

Above: The N10 with a 450-hp Napier Lion engine. The N.E.Co stood for the Australian N.E. Company.

Above: The front undercarriage struts have collapsed on the Napier Lion powered N10.

Above: With short span wings and Napier Lion engine, Lt Col Vincent Nicholl taxies the N10, now G-EALQ, for the Schneider Trophy Race held at Cowes in September 1919. The Lion was to be the engine for the nearly all the future variants of the III series.

Above: With two bay wings G-EALQ took part in the Air Ministry competition in September 1920 with this amphibian gear fitted to the floats.

Above: A poor view of G-EALQ but one that shows the position of the "G" on the tailplane.

Below: Note that the engine side panels have been removed.

for bombing and reconnaissance. Manoeuvrability was "fairly good."

Directional controllability was considered in need of improvement. The oil tank in the top plane was criticised as having only 4 gallons when at least five were required. When the oil level in the tank was getting low *as on more than one occasion* the pipe was uncovered allowing air to get into the system causing erratic oil pressure.[2]

Wing Cmdr J.N. Travers reported on testing the "Flaps" (the Fairey Patent Camber Gear) on the N10. It was found that the flaps were an advantage when leaving the water with full load. It was also necessary to use them if the pilot wanted to make a slow landing.

A very fast landing cannot be avoided in this particular machine, even in a strong wind if flaps are not made use of. If a pilot neglected to use flaps when landing at sea in the least swell, in all probability serious damage would occur to undercarriage and floats. This is very important point as the majority of pilots may in the excitement of the moment when forced to land at sea neglect the use of the flaps.

It was found that the use of the flaps when over 4,000 feet assisted climbing to a large degree. Despite this and the above comments, Travers *would not recommend flaps on this type of machine as it is seldom necessary to gain altitudes; majority of submarine patrols are carried out between 500 and 1500 ft. In the majority of heavy seaplanes without flaps on wings, a slow enough landing can be accomplished fully loaded without danger to undercarriage. Wings with flaps (that) are not used but kept in normal position are not as efficient as wings without flaps.*[3]

Testing of N10 was carried out with different loads and different position of the "wing-flaps". The landing speed except that with the heaviest loading was determined by levelling out after a glide with the engine off and noting the stalling speeds. With a loading of 12 lb/ft² the machine failed to get off in a wind of about 15 knots. The load was reduced and the test repeated until the machine left the water. This occurred when the load had been reduced by 446 lbs to give a load of 11.04 lbs/ft². The landing speed for this load was found by stalling the machine onto the water. With the three heaviest loads and the flaps over 5° from the normal, the machine was not easily controlled, and with the heaviest load the lift wires broke on landing. The results of the tests are given in the following table.[4] (see below)

Another report in February had noted the following structural defects to the N10:-

The sides of the Parnall built main floats buckled when the trolley was placed under them. As no quick release arrangement was fitted the lower part of the chassis had to be dismantled in order to remove the float for repairs.

The method of attaching the chassis cross wires was weak and two had already been broken and replaced.

The "adjustable flaps" on the upper and lower planes could be adjusted by the pilot independent of the ailerons. They were connected by two wires on either side with quick releases. These are released when the wings are folded. This was described as a difficult operation at sea. When the fittings were free the flaps dropped vertically on their hinges so that when the wings were folded the lower flaps were awash and uncontrollable.

The oil tank fitted to the centre section of the upper plane between the leading edge and front spar *now assumes a rough shape being blistered and indented.*

The aileron control cables tightened as the wings were folded causing the ailerons to droop and foul the bracing wires of the tail fin. This was thought easily to overcome.

The jury struts buckled when the wings were folded and needed to be of heavier gauge tubing.

The position of the pilot under the centre section was criticised as being difficult for him to make a quick exit, but the upper plane, being low, offered a good outlook over the plane. It was recommended that an arch be employed to stiffen the centre bay and the fairing behind the pilot be lowered to allow for a quick exit.

No bomb gear was fitted. *It cannot be seen how two 230 lbs bombs can be fitted horizontally owing to the short distance between the axles and the fact that the bombs must be under the C of G. It is suggested that they be fitted vertically between the pilot and passenger's seat.*[5]

Considerable correspondence between the Dept of Aircraft Production and the Fairey Company took place as to the strength of various members of the fuselage and wings. It

Effect of Wing-flaps on landing speed of N.10

Weight in lbs	3,470	3,880	4,365	4,850	5,266
Weight lbs/ft²	7.29	8.15	9.16	10.19	11.04
Speed in kts					
Flaps 14° down	-	-	41	-	50
Flaps 2° down	43	43	44	46	54
Flaps 5° down	44	44	45	47	-
Flaps normal	46	47	47	49	-
Flaps 5° up	47	49	49	54	
Flaps 8° up	49	51	52	-	-
Source: Report N.M.145A. 04 May 1918. TNA AIR2/969.					

Fairey N10

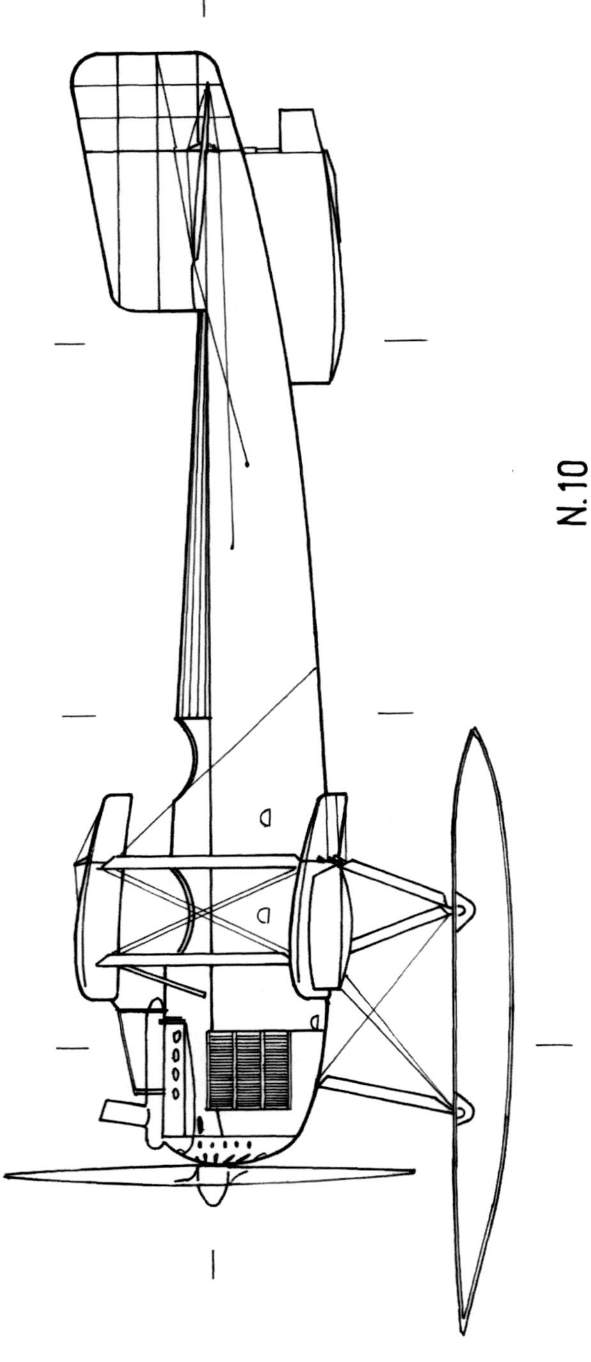

N.10

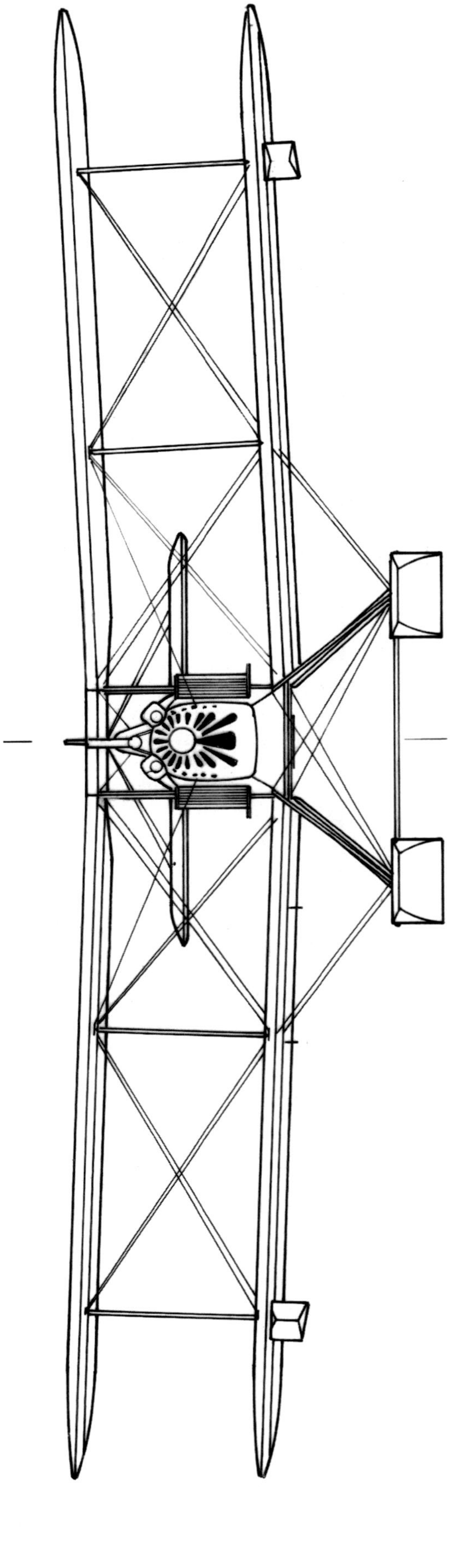

N.10

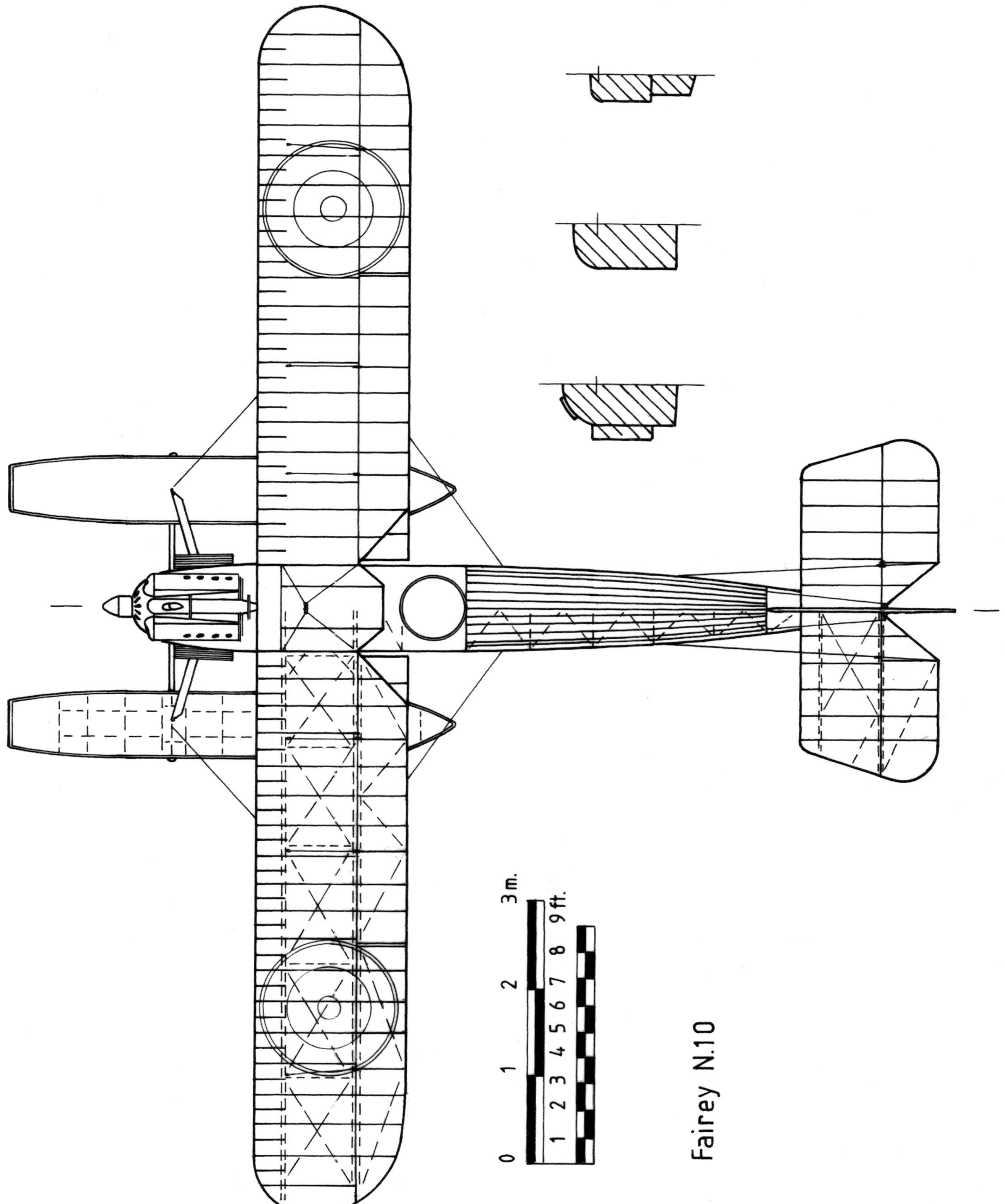
3m.
0 1 2
1 2 3 4 5 6 7 8 9 ft.
Fairey N.10

Specifications F.128/N.10	
Source	**Report N.M. 107**
Span, Upper	62 ft 8 in
Span, Lower	44 ft 8 in
Folded	13 ft 7 in
Length	37 ft 3 in
Length folded	44 ft 8 in
Height	13 ft 8 in
Span of tail	12 ft
Airscrew	
Type	Fairey 2 bladed LH
Dia	12 ft
Pitch	8 ft 2 in
Floats	
Length	19 ft
Width	2 ft 10 in
Depth Max	1 ft 11 in
C to C	9 ft
Wing Area, ft^2	570
Total Wt, lb.	4,892*
Empty (9 gall water), lb	3,235
Crew, lb	360
Lewis Gun & 3 double trays, lb	33
Scarff Ring, lb	24
2 x 230-lb bombs & gear, lb	500
52 A W/T set & lamp, lb	100
Misc Gear, lb	5
Fuel, lb	555
Oil, lb	38
Extra fuel, lb	23
Fuel, gal.	78½
Oil, gal.	4
Endurance	Approx 4½ hrs
Speed	
at sea level	
at 3,000 ft	82 kts
at 5,000 ft	80 kts
at 10,000 ft	71 kts
Height	
Max Effective	10,300 ft
Greatest reached	10,600 ft
Engine	260-hp Sunbeam No. 2436

Specifications F.128/N.10	
Source	**Report N.M. 107**
Climb to	
1,000 ft	1 min 55
2,000 ft	4 min 15
3,000 ft	6 min 45
4,000 ft	9 min 30
5,000 ft	12 min 30
6,000 ft	16 min 00
7,000 ft	20 min 00
8,000 ft	24 min 40
9,000 ft	30 min 25
10,000 ft	37 min 45

* The figures quoted are from the Report but add up to 19 lbs less than the Total Weight. This missing weight may be the Dead Weight for the extra tank that was noted as being required.
Source: Report No. N.M.107, 28.02.18. TNA AIR2/969. This is the only specification sheet for the N10 that the author has been able to discover.

appeared that these had all been resolved but for the front body strut by May 1918. However, this was not the case, as in March 1920, when the IIIC was going to be ordered into production with modifications as the IIID, it was to be checked to make sure it could take the additional load. The section that conducted the stress calculations noted that the machine had been stressed out when seaplanes were in the hands of T.1.N and *we had no power over the design or strength but simply stressed out the machine at the request of T.1.N. The latter section pleased themselves as to whether or not they adopted our recommendations.*

During reorganisation we were given full control over the strength of Sea machines as over land machines, but as this machine was then in service we did not have any queries on it till last year. We then found that several members were below the required strength and had no notifications that alterations had taken place. Once the proposed modifications were in hand the section would check the strength of the machine.[6]

Due to these concerns about the strength of the IIIC it was subject to load testing in 1921. *The machine was inverted and supported under the vertical struts in the front portion of the fuselage on beams arranged so as to give a single point of support (approximately in the plane of the centre of gravity) on each side of the fuselage. The vertical struts in the fuselage at the points of support were strengthened by means of auxiliary struts.*

Tests were conducted until June 1921. Conclusions were that the spars had sufficient strength for the assumed conditions of load but the eyebolts attaching the lift wires to the main planes were of such a form as to cause excessive deflection of the bolts under load. New eyebolts were better.[7]

It says something for the Fairey Company's manufacturing quality that there were no reports of structural failure in the IIIC seaplanes during their arduous Russian campaign.[8]

On 9 April 1918, the Fairey Company was informed that it had *now been definitely decided that no new type of float seaplane is to replace the Short 184 type seaplane as the standard short distance submarine patrol machine.*

After much discussion it was agreed that the float seaplane of a medium size is not sufficiently seaworthy to carry out coastal patrol in ordinary weather, the work being carried out in future by large open sea flying boats and aeroplanes.

You are therefore requested to discontinue all work on machine No. N.10 fitted with to fulfil the N.2.B. Specification and concentrate your designing staff and experimental shops on the N.10 (N.2.A. Ship Aeroplane) now in production and the N.4 Type large Flying Boats.[9]

At the 20th Progress Committee Meeting it was reported that the "Fairey Type N.2A seaplane with 260 h.p. Sunbeam" had carried out satisfactory trials. It was now proposed *to fit the machine with a land chassis and carry out further trials, with a view to the type being of use as a two-seater for ship use.* The question was asked whether the type could be ordered if the trials were satisfactory or would it be better to order a small number, such as three or six experimental machines for further trials before placing a large production order. It was decided to place an order for six experimental machines.[10] It is thought that this order was for the aircraft that received the serials N76 to N81.

After testing as a seaplane, the N10 was modified to a landplane as the Fairey IIIA with a frontal radiator. It was later to be reconverted to a seaplane with Type IIIB wings.

The Fairey IIIA

The N10 in landplane configuration was out for engine trials on 20 November 1917, and *should be ready for despatch to Grain this week.*[11] It reached the Grain Test Depot on 5 December.

The Fairey IIIA was ordered into production with 50 (N2850 – N2899) being contracted from Fairey for the RNAS to replace the Sopwith 1½ Strutter. The observer had a single Lewis gun on a Scarff ring and the aircraft could carry a load of bombs on external racks but no bomb load has been identified. An order for 50 "Fairey IIIA N.10 Seaplanes N.2.A" had been received by Fairey and the Company wanted to use U-bolts in the fuselage construction.[12] (The inclusion of "Seaplanes" in the report is considered a clerical error as the IIIA was a landplane). Initially this request was granted, but it was soon pointed out that *this permission* (is) *rather unwise in face of the trouble with the 320 Sunbeam Shorts. Although the cause of the fuselage failure on these machines was never discovered definitely it is at least significant that the three machines with 'U' bolt fittings gave trouble, whereas later machines with plate fittings have been O.K.*[13] Permission was then withdrawn.

The first production IIIA was flown by Lt-Col G.L.P. Henderson, of whom we shall see more of post-Armistice, at Northolt on 1 June 1918. In August, the first Fairey IIIA, N2850 (Sunbeam Maori) had completed its stick and unstick tests at Martlesham Heath and was to go to Grain.[14]

The IIIA type was to be delivered with normal wheel or skid undercarriages for deck landing experiments. However, in *view of the unsatisfactory reports on the Skid Chassis on the Fairey Type 3.A Machine, it is considered better to fit the wheel type chassis.* In reply it was stated that the requisition had not yet been received but that action had been initiated to obtain the necessary chassis. It was later reported that 50 wheel chassis had been ordered from Fairey and the first was expected to be delivered on 10 July 1918, with three per week thereafter. The first machine to be fitted was to be N2889 which was to be delivered to Renfrew on 20 July if rail could be obtained. All the machines up to N2888 were delivered with skid undercarriage. It was anticipated that there would be a surplus of wheel chassis of three per week that would be sent to Renfrew for fitment to machines already delivered.[15]

The Fairey Ship Aeroplane was the subject of a request from Harry Busteed asking how the flotation gear was to be arranged. It was noted in reply that the hydrovane was that designed by the company and seemed to be satisfactory. It was thought that it *might be wise to send one of the Fairey machines to Grain* to test the hydrovane gear.[16] N2863 was fitted with a hydrovane. Details of flotation testing of the IIIA have not been discovered to date.

Richard Fairey wrote to the Technical Department Aircraft Production on 3 September 1918, recording that the company had *recently weighed some of the rebuilt Maori II engines now being supplied to us for use with Type III B Seaplanes, and enclose a comparison of these with that as fitted in the original Type III B machine Number N.10. You will note it is approximately 105 lbs in excess, the bulk of which is due to the engine itself and certain extra accessories which have since been found necessary.*

Fairey made the point that he had frequently complained that the Maori engine was extremely heavy per horsepower, however this time he was complaining about the Martlesham Heath comparative testing of the IIIA, with its heavy engine, with the Parnall Panther ship plane with the B.R.2 rotary engine, *one of the lightest possible* engines. He wanted a rebate of weight for the type trials corresponding to the surplus engine weight over that specified.[17] In the first six months of 1919 it had been proposed to replace the Sopwith 1½ Strutter and the Fairey IIIA with the B.R.2 powered Parnall Panther.[18]

The Fairey IIIA went into production as a ship plane but the war ended before it could prove its worth. As noted below, the Navy did not fit the type into any specific role and all but three were sent directly into store at the Renfrew Aircraft Acceptance Park. The IIIA aeroplanes were seen as possible replacements for the aging D.H.6 biplanes used for inshore anti-submarine patrols. Only N2850 had arrived at Luce Bay for No. 523

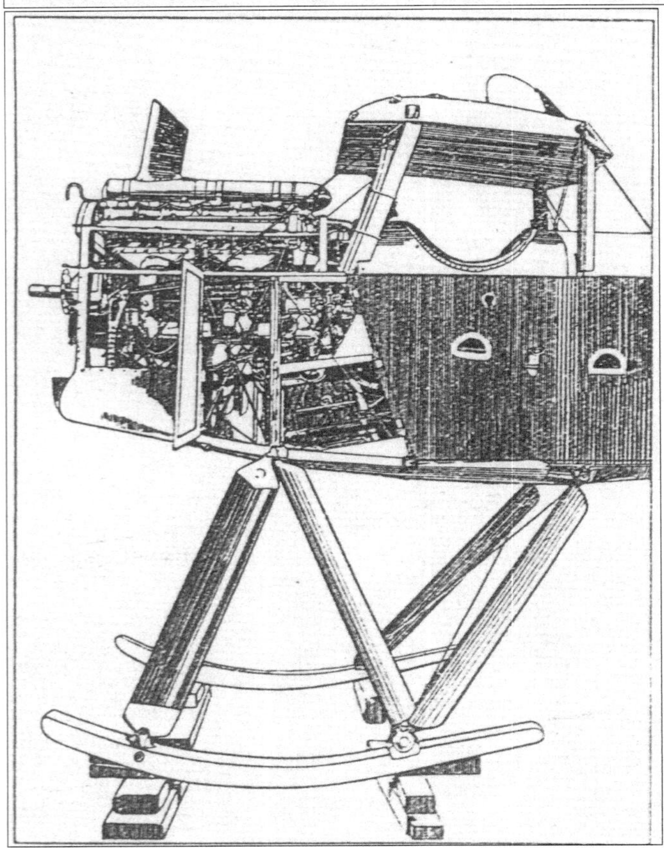

Above: Forward portion of the fuselage with one tank missing. 250-hp Sunbeam Maori engine.

Left: Skid gear on Fairey IIIA..

Facing Page, Above: Rear section of fuselage of IIIA showing tail skid & air bags.

Facing Page, Middle: Tail section showing tail skid and tail bracing.

Facing Page, Below: Instrument panel of Fairey Seaplane Type III 250-hp Sunbeam Maori engine. Compare with photographs.

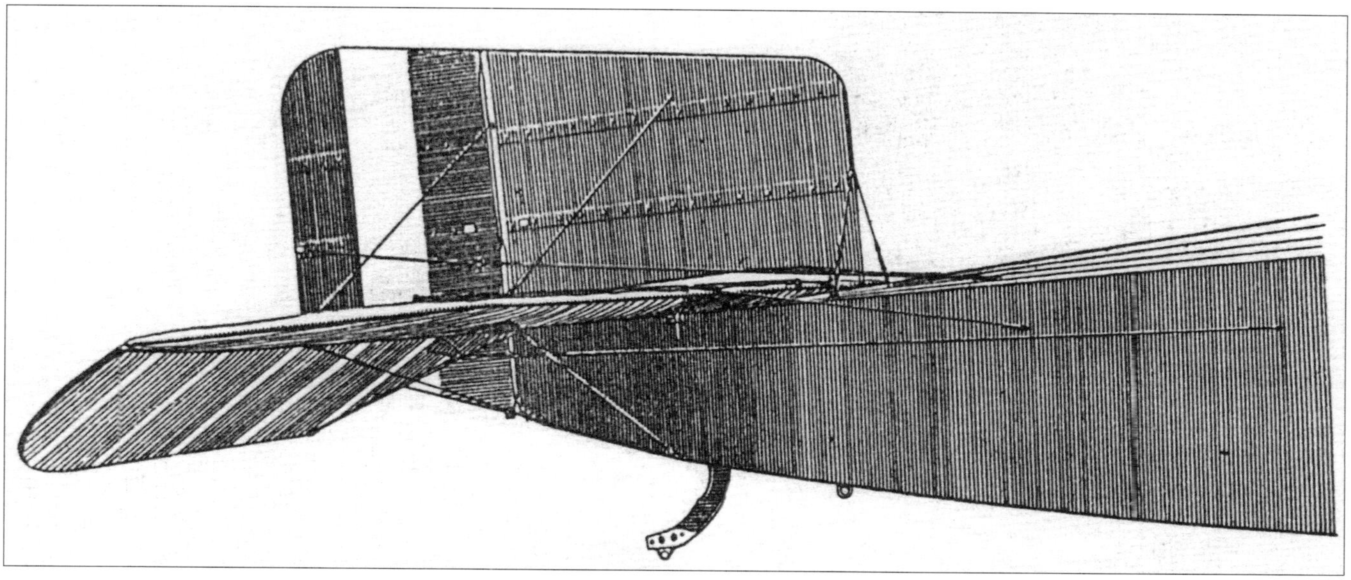

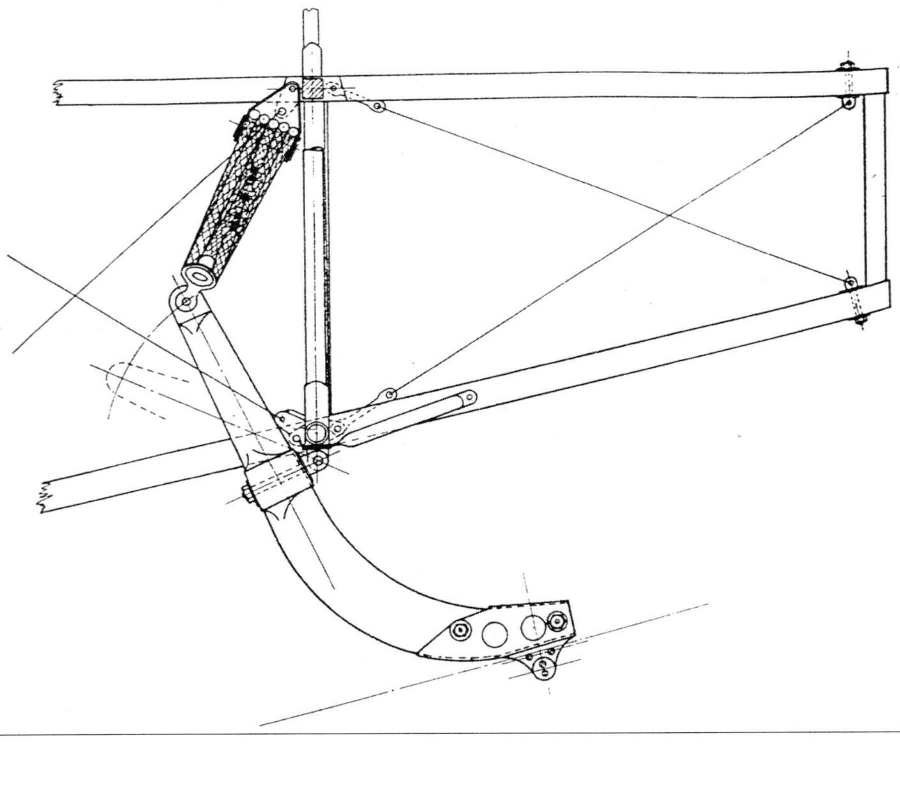

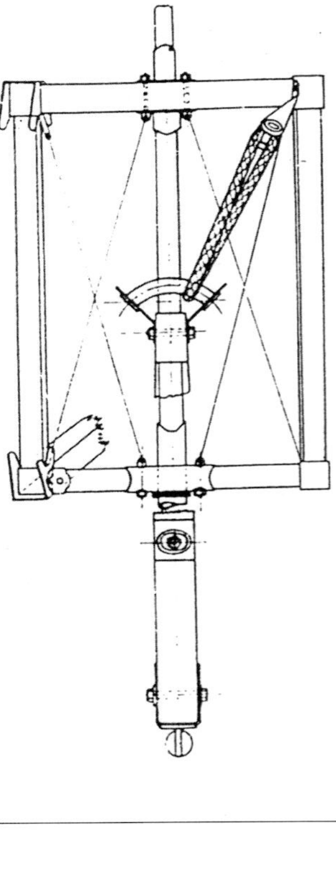

Above: Rear view of a Fairey IIIA. Note the cutouts in the centre section to allow the wings to fold.

Above: Preparing to swing the prop of a Fairey IIIA.

Facing Page, Above: Sketch of the main plane strut fitting and quick release.

Facing Page, Below: Tail skid for land version.

Flight before the Armistice altered plans.

N2876 was purchased by the Navarro Aviation Company and received a temporary registration on 6 June 1919. This machine was given the registration G-EADZ on the introduction of the permanent register. Re-registered as G-EAMY after its sale in September 1919 to Lt-Col G.P.L. Henderson. He had it converted to IIIC civil standard and took it to Sweden.

IIIA Shipborne Reconnaissance Tractor Biplanes		
Serial Allocation	**Contract**	**Notes**
N2850 – N2899	No. AS3 35160 (Br 257)	Proposed N2853 -N2862 skid u/carriage; N2864 – N2874 skid u/carriage & air-bags; N2889 – N2899 wheel u/carriage.

Above: The first IIIA N2850 has had the background removed from the photograph for use in an identification chart.

Above: Rear view of a Fairey IIIA. Note the cutouts in the centre section to allow the wings to fold.

Facing Page, Middle: Fairey IIIA shipboard reconnaissance tractor No. N2851 at Grain on 18 May 1918. This photograph was used to illustrate the Fairey IIIA in the *Types of British Seaplanes, Flying Boats and Ship's Aeroplanes* manual.

Facing Page, Below: N2851, at Grain on 18 May 1918. Most of the type were stored and never entered service. N2851 was second of a batch of 50 that had various undercarriages. N2853 to N2874 has skid undercarriages but these proved unsatisfactory and wheel undercarriages were substituted.

Above: N2851 was delivered to the Isle of Grain the W/E 18 May 1918. This is the incident of 17 May 1918, when it was crashed on its nose. On 14 July, the repaired aircraft was taken by Capt Arthur Leslie Simms, DSC, from the Isle of Grain air station on a delivery flight to Hylton. He overshot and tried to gain height but crashed. The 20-year old Simms died of his injuries.

Above: Skeleton view of the fuselage showing construction.

Facing Page, Below: Installation and controls of the Rolls-Royce Eagle VIII engine in the IIIC. Before fitting the engine, the standard feet for tubular bearers were removed and a special plate fitting attached. The brackets on these plate fittings were then bolted to the engine bearers. For a left-hand tractor the engine centre-line was inclined 1° towards the left, and similarly 1° right for a right-hand tractor.

Above: Fuselage of N9241 showing engine bearers and main fuel tank. Note the seat in the gunner's cockpit. N9241 was ordered as an IIIB but delivered in the W/E 30 January 1919, as an IIIC. It was delivered to North Russia by HMS *Nairana*.

Above: Main plane of the IIIC.

Left: Petrol and air-pressure cocks in the IIIC pilot's cockpit. On the starboard side of the cockpit and aft of the control cocks were ten pipes that were connected with the petrol control system, that were placed one above the other. The forward group was called the "petrol assembly" controlling the fuel feed, and the after group the "air pressure assembly" controlling the air pressure to the fuel tanks.

Facing Page, Above Right: The gunner's cockpit of the IIIC showing the No. 11 Scarff ring and wireless gear. Note the open floor panel. A standard Lewis gun was the observer's armament. The gun could fire directly over the centre-section or over the tailplane. Stops were provided so that the gun could not inadvertently fire on any part of the aircraft.

Left: Pilot's cockpit of the IIIC showing the instrument panel. A Thompson Bennett type hand-starting magneto was fitted to the top starboard corner of the instrument panel. The ASI is in the centre of the dashboard. The indicator plate for the Camber gear is on the port side of the panel.

Above: Side radiator and shutters of the IIIC.

Right: Engine cowling, reserve water tank, radiator and shutter of the IIIC. Note the machine gun and blast tube. H.F. King noted that although a Vickers gun for the pilot had been reported in the Fairey IIIC "no such installation can be identified in photographs" (*Armament of British Aircraft 1909–1939*, Putnam, 1971. P.149).

Above: View showing starting gear and magneto. Note also the radiator and screen.

Left: View of the oil tank, starting gear and the reservoir for the CC gear for the synchronized Vickers gun.

Specifications Type IIIA

Source	1	2	1	3	4	5	6
Type	III[3]	III[3]	IIIA	N2850 IIIA	IIIA	IIIA	IIIA[4]
Span	46 ft 2 in	46 ft 1 ½ in	46 ft 2in	45 ft 10¼ in	44 ft	45 ft 10 in	
Length	36 ft	35 ft 0 in	31 ft	31 ft 0 in	31 ft	31 ft 0 in	
Height	11 ft 10 in	11 ft 8 in	10 ft 8 in	12 ft 0 in	10 ft 8 in	12 ft 0 in	
Chord	5 ft 6 in	5 ft 6 in	5 ft 6 in		5 ft 6 in		
Gap	5 ft 7 in		5 ft 7 in		5 ft 7 in		
Span of tail	13 ft		13 ft		13 ft		
Incidence		4½°					
Dihedral		2°					
Airscrew Diameter				12.0 ft		AB4475M	
Airscrew Pitch				2890 mm[1]			
Areas in ft²							
Wings	476		476	464.5	476	464.5	464.5
Ailerons	66		66		66		
Tailplane	34.2		34.2		34.2		
Elevators	34.2		34.2		34.2		
Fin	9.8		9.8		9.8		
Rudder	9.8		9.8		9.8		
Weights in lbs							
Empty				2,690[2]		2,690	2,690
Crew				360		360	360
Petrol				406 (56½ gall)			
Oil				40 (4 gall)			
Fuel & Oil						446	446
Military load				449		449	449
Total				3,945	3,694	3,945	3,945
Fuel Capacity	70 gall		70 gall		76 gall		
Endurance					4½ hrs		
Speed at sea level					95 kts		
Speed at 10,000 ft				97 mph	90 kts	97 mph	84 kts
Speed at 12,000 ft				93 mph			
Landing speed					40 mph		
Climb to 2,000 ft							3.6 min
Climb to 5,000 ft				10.0 min	7.10 min		
Climb to 6,500							14.0 min
Climb to 10,000 ft				26.0 min	17.30 min	26 min	26.0 min
Service Ceiling				13,300 ft		13,500 ft	
Engine		250 Sunbeam				Sunbeam Marci	

Notes: 1. Pitch was marked as 9.35 ft but measured as 2890 mm. 2. Empty weight included water.
3. It appears that the N9 & N10 were considered as the Fairey III, however this has not been confirmed.
4. Report M.220 of July 1918.

Sources: (for Type IIIA Specifications)

1. Bruce, J.M. *British Aeroplanes 1914–1918*, Putnam, UK, 1957.
2. *Technical Notes Fairey Seaplane Type III*. Although this manual states it is for the Type III seaplane with 250-hp Sunbeam engine, and the rigging of the floats is included in the instructions, all the drawings are of the landplane with a skid undercarriage, the IIIA.
3. H Busteed notes on Fairey Ships Aeroplane N2850. Copy in RAF Museum J.M Bruce Collection Box 26.
4. *Jane's All the World's Aircraft 1919*.
5. Chart – "Record of Performance of British Aeroplanes," May to Sept 1918, Sheet No. 8. Date of Trial Report July 1918. Report M220. TNA AIR1/708/27/11/03.
6. "Table of Performance of Seaplanes & Ship Aeroplanes." August 26, 1918. TNA AIR1/708/27/11/431.

Above: An anonymous Fairey IIIA. The round object next to the fuselage may be a field refueling tank.

The Fairey IIIB

The IIIA was followed by the IIIB seaplane that was designed to take the N.2.B bombing load. While it was similar to the N10, with the same fuselage and tailplane as the IIIA, the IIIB had larger floats and vertical tail surfaces. The wing area was increased by giving the upper wing large king-post braced extensions in order to take the N.2.B specified load. The observer retained the gun ring in the recessed mounting of the previous type III aircraft. Bombs were carried on tubular frames under the fuselage, a typical load being two 230-lb bombs in tandem.

The calculations for the machine fitted with the land undercarriage found that the centre-section front and rear spars, as well as the front spar of the inner bay, were under strength.[19] As before the wings could be folded and Fairey Patent Camber Gear was fitted. Ailerons were fitted to the upper wing only. A frontal radiator was fitted and the Sunbeam Maori II was again the power plant. Drawings of the IIIB were reported as almost complete on 21 November 1917.[20] Instructions to proceed on the contract for the 30 IIIB seaplanes was given on 9 May 1918.

Twenty IIIB seaplanes had been delivered to the Hamble Marine Acceptance Depot before the armistice. The type started to reach units before the Armistice; Nos. 219 Squadron at Westgate in August, No. 230 at Felixstowe in October, and No.229 Squadron in Flanders in November. In January 1919, No. 229 moved to Great Yarmouth from where it undertook

Right: The frontal radiator the IIIB consisted of a standard block of 7 mm x 100 mm tubes, 3 ft 8 in high, carried in front of the engine by mild steel tubular stays. The radiator was fitted with a filling tube at the top and with two inlets and one outlet connection. The reserve water tank is in the upper portion of the radiator. This reserve makes up for any water lost by evaporation. When empty the radiator weighed 75 lbs, and the minimum capacity of the system, including radiator, engine and piping was 10 gallons and the maximum capacity was 12 gallons.

Below: Skeleton view of mainplane of the IIIB. The fuselage was of conventional construction divided into 13 bays

Above: Pilot's cockpit instrument panel.

Below: Front view of an IIIB.

Above: Lt Droton leaning on float of a Fairey IIIB at Westgate, March 1919. Note that a large numeral or letter is painted on the fuselage aft of the cockade.

Above: The serial of this IIIB cannot be discerned. The engine panels were probably navy grey with tar-black floats. Note the profusion of stencils on the interplane struts and wingtip floats.

Above: Another view of the IIIB whose serial cannot be discerned. The engine panels were probably navy grey with tar-black floats. Note the profusion of stencils on the interplane struts and wingtip floats.

Above & facing Page, Above: Fairey IIIB N2230 at Grain 19 October 1918.

Above: Maori II powered N2245 on the beach at Westgate. Delivered in the W/E 17 October 1918, this seaplane served at Westgate and Yarmouth.

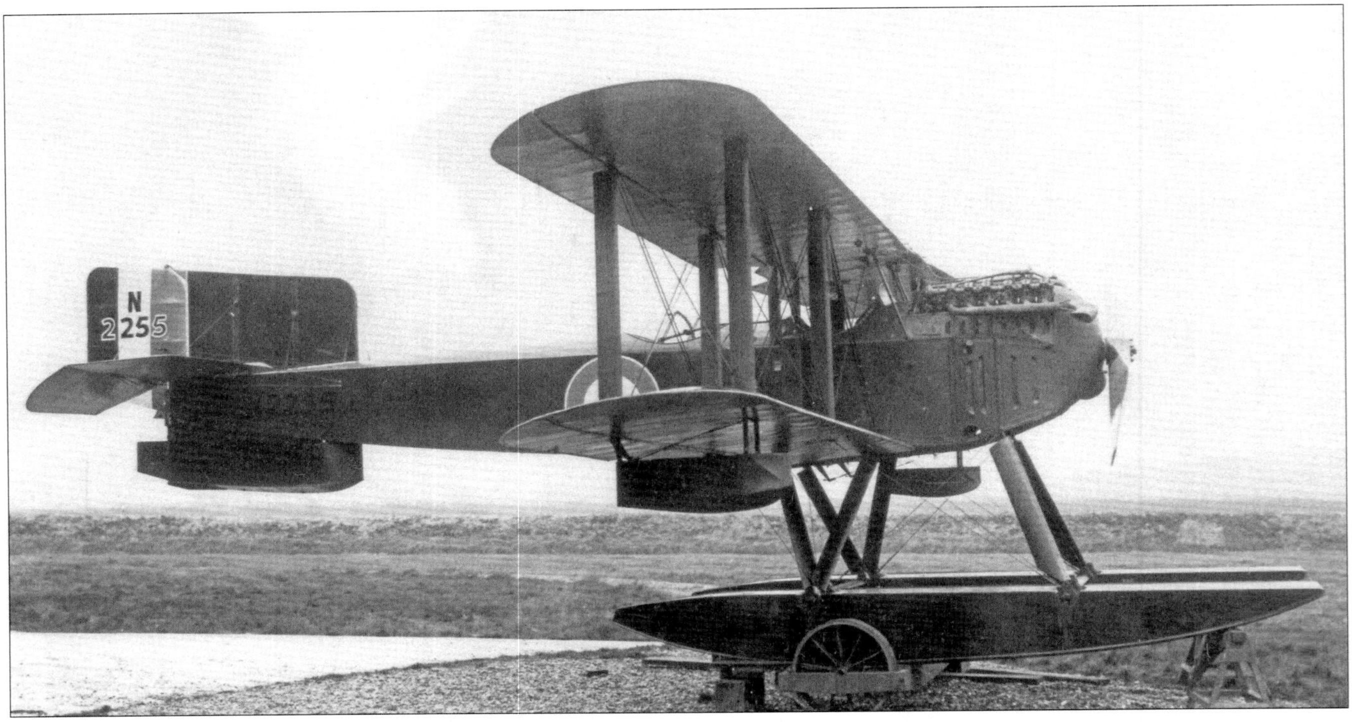

Above & Below: This IIIB was delivered with a 375-hp Rolls-Royce Eagle VIII, hence being to IIIC standard. At Fairey's Hayes facility on 2 November 1918, it underwent trials around the end of January 1919. It was registered as G-EAPV and ended up in Sweden.

Facing Page, Above: Fairey IIIB in flight. Possibly Serial N2243. Note "A" on fuselage side and fuselage top decking. No. 219 Squadron, 1919.

Facing Page, Below: Maj Livock flying the Fairey IIIB from Westgate. Most probably the same as the above. The rear seat occupant is holding onto his hat.

A
A

A
A

Above: Launching Fairey IIIB at Westgate, Maj Livock pilot.

Above: Fairey IIIB at Westgate.

Above: This is the same Fairey IIIB. The "A" can just be made out on the fuselage side.

Below: Fairey IIIB at Westgate with Short 184 in background.

Type IIIB Specifications

Source	J.M. Bruce Data	Janes 1919	IIIB/IIIC Manual
Span upper	62 ft 8 15/16 in	62 ft 9 in	62 ft 9 in
Lower	44 ft 8 7/16 in	44 ft 8 7/16 in	44 ft 8½ in
Length	37 ft 1 in	36 ft	36 ft 0¾ in
Length wings folded			43 ft 3½ in
Height	14 ft	13 ft	13 ft 6¾ in
Chord	5 ft 6 in	5 ft 6 in	5 ft 6 in
Gap	5 ft 7 in	5 ft 7 in	5 ft 7 in
Span of tail		13 ft	13 ft 0 in
Dihedral	1° 30’	1° 30’	1° 30’
Incidence	4° 30’	4° 30’	
Floats			
Length			17 ft 10 in
Maz width			2 ft 10 in
Max depth			1 ft 8½ in
Areas in sq ft			
Wings	570	570	570
Ailerons	46	46	46
Tailplane	34.2	34.2	34.2
Elevators	34.2	34.2	22
Fin	12.4	12.4	
Rudder	12.4	12.4	15.3* excluding balance portion
Total Wt, lb		4,892	5,083
Light Wt, lb			3,423
Military load, lb			690
Crew Wt, lb			360
Fuel Wt, lb			562
Oil Wt, lb			48
Fuel Capacity, gall.	76	76	77
Endurance		4½ hrs	4½
Speed at sea level		83 kts	
Speed at 10,000 ft		76 kts	
Landing speed		42 mph	
Climb to			
5,000 ft		12.30 min	
10,000 ft		36 min	
Min safe flying speed			36 kts
Min safe gliding speed			42 kts

IIIB Serials		
Serial Allocation	**Contract**	**Notes**
N2230 – N2259	Nos. 38a/392/C374 & A.S.16660 (BR475)	From N2246 most converted to IIIC with 375-hp Rolls-Royce engine.
N9230 – N9259	No. 38a/513/C522 & A.S.22394/18 (BR522)	Ordered as IIIB but delivered as IIIC.

patrols of the Norfolk coast. The end of hostilities ended the Fairey IIIB's career before it had really started. The few that entered service were used for spotting mines post-Armistice.

The Fairey Fighting Seaplane

In May 1918, a Meeting was held to discuss whether a Fighting Boat Seaplane should be developed for escort duties with large flying boats. The Curtiss and Felixstowe flying boats were having to put up with increased opposition in the North Sea from the Brandenburg fighting biplanes that were operating from Zeebrügge and it was predicted that up to five boats would have to be sent in order that they could satisfactorily carry out their reconnaissance of the German occupied coast.[21]

The following month another report came about due to the *increased enemy activity of H.A. in the North Sea.* Yarmouth reported that it had to send out a formation of six Felixstowe F.2a flying boats when a patrol "to the mouth of the Ems and of the Dutch Islands down to the Hake Light Vessel" was required. Felixstowe station also reported that the greatly increased enemy air activity caused increasing difficulty for patrols from this station, it being necessary for the shortest patrols to be of three machines as hostile aircraft were liable to be encountered only 20 miles from the English coast. *There is an urgent demand for a high-performance seaplane to operate from Felixstowe to engage H.A.*

So far, the enemy seaplanes had only operated in favourable weather conditions *and there appears to be every reason to believe that they are not more seaworthy than any small seaplane that has been used up to the present.* It was therefore concluded that the good seaworthiness of a seaplane to oppose them was therefore not essential as *it will only be required on comparatively calm days.* It was pointed out that De Havilland D.H.4 land machines now accompanied the flying boats as much as possible and there was a definite need for the D.H.4 aircraft to be replaced by a seaplane as continual flying over water in a landplane placed *a severe strain on the pilots and their work would be carried out more effectively if they could feel less dependent on their engine for safety.*

It was suggested that there were *in existence a certain number of Fairey Type IIIA Ship-planes fitted with skids and flotation gear. As far as can be seen these machines are not required for Service in their present form and it is suggested that they should be converted to seaplanes and fitted with Rolls Royce Series 8 engines and additional fuel carried to give them 4 ½ hours duration at full power at sea level.* With the pilot's synchronised machine gun and the observer having a Lewis gun with five double trays of ammunition, the type was thought to have the requirements to accompany the Large America flying boats. The memorandum concluded with the remark that the boats were consistently overloaded and that this overloading was *very largely responsible for the present unpopularity of this type of boat at Yarmouth.*[22]

August saw a conference at the Air Ministry that considered whether the present policy of concentrating on the boat type for anti-submarine work was justified, or whether new developments of the floatplane would meet requirements more effectively. *Until experiments with the Fairey 3b fitted with Rolls Royce engine have been conducted it cannot be said that a limit in the development of the single engine Float Seaplane has been reached.* The conference also considered that the *present project of fitting D.H.9's with floats and floatation appliances should proceed also for offensive work at immediate distances.*[23] The Air Ministry considered that modifications to the flying boats would make them more capable of self-defence. *It is proposed to test these on service in the North Sea. It is feared that their fuel range will prevent them from operation to the limits at present covered by the Boat type Seaplanes, but their speed and increased handiness may give them sufficient fighting value to meet German Seaplanes at intermediate distances.* It was proposed to reduce the number of flying boats supplied to Felixstowe and Yarmouth by three on the supplying of the Fairey machines. In addition, it was proposed to convert three De Havilland D.H.9 biplanes into floatplanes, and three other D.H.9 machines with emergency flotation gear, for the purpose of comparative trials.

The experiments referred to were to lead to the Fairey IIIC. Following Maj Wright's report on Felixstowe and Yarmouth, the matter of fitting an Eagle VIII to the Type III was then considered with Mr Fairey. With the loads specified by Maj Wright, *Mr Fairey estimates that the type III machine with the original wings, (not the cut down ship aeroplane wings which would be too heavily loaded,) should have a top speed of between 95 and 100 knots or 109 to 114 m.p.h.*[24] Fairey suggested that the most rational way to obtain the required test results would be to take a fuselage from the present production line and fit it with an Eagle engine. This fuselage, with two sets of wings, Type III and IIIB, would then be sent to Hamble and tested with the load required. *As all this gear is practically made, Mr Fairey estimates that a test machine could be ready in about three weeks from the receipt of the engine.* It was submitted that this should be done at once as *the condition of affairs is really that the enemy have beaten this country in the design of a fighting seaplane.*[25]

Fairey's proposed was agreed to as a note, dated 16 August, stated that the engines should be left hand tractors and the first engine should be sent to Faireys as soon as possible. It was decided that six Type III or IIIB seaplanes should be fitted with the Rolls-Royce engine and guns and be tried at Yarmouth *against the Hostile Aircraft at present operating near that station.* With the proposed load of 1,555 lbs (including 120 gallons of fuel), the machine was carrying an increase of 388 lbs over that of the original N10.[26] It was considered that a load of 120 gallons of fuel was too great and while *doubtless the machine will fly with this load it is really asking rather much of it, and that performance will suffer to too great an extent. Also, if the load is increased too much the question of structural strength will have to be re-considered and Modification made to the Planes, as this machine has been designed very closely to the required factors.* It was suggested that 100 gallons was a more suitable fuel load.[27] The following month a file note referred to the machine as the Fairey Type C.3, later corrected to 3.C.

The Fairey IIIC

Above: N2246 was laid down as an IIIB but was chosen for conversion to the Fairey IIIC with a 375-hp Rolls-Royce Eagle VIII engine, the first conversion. Completed in September 1918, it was delivered to the Isle of Grain on 18 October. It was used extensively for experimentation and was converted to a landplane in June 1920, and performed deck landings on HMS *Eagle* that July.

Instructions to proceed on the contract for 30 IIIC seaplanes (N9230–N9259) was given on 5 July 1918.[28] Problems soon arose with obtaining the French manufactured Sacana self-sealing petrol tanks that had been ordered for the IIIC in place of the M.I.D. tanks *built in this country*. The French tanks had been ordered as it had been understood that an early delivery would be given. [29] The company was not completely to blame as modifications that were *not absolutely necessary* but were desirable were requested to be incorporated in the tanks *now in hand.*[30] The file shows the usual blame shifting from one Department to another, however, *the whole set of 12 tanks are promised for Wednesday next the 9th* October.[31] The fiasco did not end here with a letter from Paris revealing that *you have sent a drawing of the extra tank behind the Pilot's seat. When I was in London I was told that this tank was not required and so only the main and bottom tank rear tank was ordered. I have told the Sacana Company to get out the model for the top rear tank and unless you telephone us to the contrary, these tanks will be delivered with the others making a set of three per machine.*[32] In late December it was considered that the order for the tanks should be cancelled.

The Test Depot checked the capacities of the tanks on N2246 and reported that their capacities were as follows:

Above: N2255 was laid down as an IIIB but completed as an IIIC in early November 1918. Became G-EAPV and ended up in Sweden. This photograph has penciled on the rear "School machine." There is no indication in the file why this should be so as the Fairey IIIC was always meant to be a fighting machine.

Above: N9230 shows the white serial application adopted for these Fairey IIIC seaplanes. The batch of 30 (N9230 - N9259) was ordered as IIIB type, but delivered as IIIC type. This batch saw the most action in the Russian intervention.

Above: The unlucky N9235 with another IIIC on Lake Onega in Russia. N9235 was shot down by AA fire on 30 Jun 1919, coming down in trees to a complete wreck. The crew were able to walk to safety.

Below: Aerial view of the RAF station on Lake Onega, showing the seaplane train, two slipways. Note the two aeroplanes on the shore, one under the gantry by the Eastern slipway. The gantry is in the water for salving aircraft. Platforms are in the process of construction. *On the left is the camber from which the Lake Flotilla operate.*

Above: Hoisting an IIIC aboard post-war.

Summary of Flights carried out by all RAF Units in Syren	
Dates	**Average Number of Effective Aeroplanes Employed**
6 June to 6 August	3 seaplanes
6 August to 6 September	5 seaplanes & 3 landplanes
6 September to 20 September	7 seaplanes & 3 landplanes.
No. service flights	403
Hours flown on service flights	755 hours 50 minutes
No communication flights	43
Hours flown on communication flights	80- hours 20 minutes
Instructional flying	44 hours 20 minutes
Test flying	82 hours 30 minutes
Total number bombs dropped	1,114
Weight of bombs dropped	Approx 30 tons
No. gas bombs dropped.	280
Machine gun ammunition fired	59, 500 rounds
Photographic plates exposed	1,395
No. prints issued	11,553
Area photographed	Aprox 253 miles
No. propaganda leaflets dropped	36,000

Source: Lt Col R.J. Bone's Report to General Officer Commanding North Russia Expeditionary Force. 1 October 1919. TNA ADM137/1713.

Front tank	48 galls
Rear top tank	29 galls
Rear bottom tank	34 galls
Total	111 galls[33]

The radiators were considered too small and rigidly fixed. Their area was to be increased and a method of controlling the air flow through the radiator was to be adopted for the production version. *No objection is made to the equipment of one machine with the radiator shown for experimental purposes, but it is pointed out that for service conditions and summer use the width of the tube block must be increased to at least 9½" or an equivalent increased frontage provided, and even in that case it may be found necessary to carry the radiators further out from the fuselage.*[34] N2246 underwent temperature trials with a Short type radiator manufactured by Motor Radiator Manufacturing Co Ltd, on 23 October 1918. The cooling surface was 148 ft^2, with a capacity of 4 gallons. The report stated that the radiator shutters were found to be quite efficient and kept the engine warm on the glide in low temperatures.

The radiator tests showed that for an English summer conditions that the radiators fitted (9" x 2' .9¼") were slightly too small unless the machine could be climbed at 80 knots in service. It was pointed out that the radiators should be 9½" wide and the supports arranged to give flexibility. *For the temperature conditions likely to be encountered in tropical Africa, it is considered that it will be necessary to increase the width of the radiators to 13".*[35]

In January 1919, a honeycomb radiator fitted IIIC machine was to be tested at Grain. Details of this experiment are still to be discovered.

Difficulty was experienced with filling the fuel tanks and they were 6 gallons short of the expected volume. It was thought that probably a small air lock in the tanks prevented the complete filling of the tanks. *This is particularly difficult to avoid in the case of flat-topped tanks.*[36] Later it was discovered

Above: Fairey IIIC N9239 undergoing maintenance at Lake Onega.

Above: Salvaging N9242. Note the white band that has been added to the rear fuselage.

Above: Fairey IIIC seaplanes lined up on Lake Onega. N9235 with N9237 and N9234 behind.

Above: Short 184 and Fairey IIIC seaplanes on the shore with what appears to be a recovery barge on the water.

Above: N9242 on the sandy shore of the lake. The conditions were ideal for operating seaplanes. Note the large bomb in the foreground.

Facing Page, Above: Short 184 and Fairey IIIC floatplanes on shore at Lake Onega. Note Bessonneau hangars and the thick forest behind the encampment.

Facing Page, Below: Gantry erected to salve a machine which smashed its under-carriage. The raft on which the fuselage was floated ashore is seen in the background. It was constructed by using Fairey IIIC main floats.

Above: N9244 on the water. There is a man on the starboard float as well as the crew in their cockpits.

Above: Another view. The writer of the report took pains to ensure that his superiors knew that the men in the foreground who are not working do not belong to the Royal Air Force. The temperature must have been mild judging by the lack of wading clothes on some of the personnel.

that the baffle plates in the tank were not drilled causing an air lock in the tank.

Maj M Wright[37] noted that six Type III machines fitted with the Eagle VIII engine had been recommended to meet the following specification: -

Pilot & passenger	360 lbs
Petrol 120 gals	870 lbs
Oil 7 gals	66 lbs
Lewis gun, Ring & Ammunition	75 lbs
Vickers (synchronised) gun & 850 rounds ammunition	90 lbs
Total Useful Load	1461 lbs

The first such machine had undergone preliminary trials with the above useful load, and was shown to have the following approximate performance: -

Climb to 10,000 ft in 17 min 30 sec.
Top speed at sea level 95 knots.
Landing speed 44 knots.

These figures were only an indication as the weather at the time of the trials was bad. The manoeuvrability was stated to be "extremely good." A version with the Liberty engine was expected to show a top speed of 102 knots at sea level. The aircraft was to be delivered to Grain for official trials and then to Felixstowe. *It is expected that the remaining five machines will be delivered to Felixstowe within the next month* (the report was dated 2 October) *and that they will be then tried on active service and definite information gained as to the suitability of the type for fighting requirements.* If satisfactory they would be utilised at Westgate, Felixstowe, Dover and Yarmouth. No great difficulty was foreseen in fitting the Liberty engine in place of the Rolls Royce.[38] The end of the war terminated the Liberty engine proposals.

Such was the haste in the desire to get the machine into service that N2246 was sent to the Isle of Grain experimental station in September 1918 without an airscrew. N2246 was tested at Grain on 23 and 24 October 1918. The results of the test report are summarised hereunder: -

Above: A salvaged seaplane on the pier, possibly that shown above. Note how everything was constructed of wood.

On the water manoeuvrability is very good indeed and machine can be turned about in a very small space. Machine sinks rather low by the tail and even in calm waters the elevator gets water over it when machine is at rests unless elevator is held up. Machine takes off quickly and easily. In the air it is stable laterally and turns slightly to the left directionally. Fore & Aft it can be trimmed by a very neat, effective and easily operated sandow adjustment which can be moved to different positions on a ratchet up and down the control column. Lateral control is quick and light. Directional control is good but might be improved and by decreasing the fin surface very slightly.

Machine is easy to land but unless the elevator controls is held hard back after landing the machine porposies between the speeds of 30–15 knots.

The field of view for the pilot is excellent forward, forward and downwards, and forward and upwards and is therefore very good for landing. The obstruction to view caused by the top plane is

Above: N9242 moving off with two bombs under the wings. Note the white band applied to the rear fuselage. N9244 in the background.

thought to be serious in a flight as when circling around another machine at the same height and in a steep bank, the enemy machine is entirely lost behind the top plane. This fault would be overcome by cutting an oval hole in top plane just above pilot's head.

The compass which is now fitted under the top plane just behind the leading edge in a streamline cowl cannot be seen when flying into the sun or strong light and it is suggested that it be fitted in the centre of the cowling to the pilot's cockpit. Just behind the cross bracing wires of centre section. The wind screen could be removed in front of the wires.[39]

The floats were 17 ft 10 in long, by 2 ft 10 in breadth and had a maximum depth of 1 ft 9 in. They were 8 ft 4 in apart, centre to centre.

In January 1919 it was reported that N2246 had been fitted with balanced ailerons. The front fuel tank had buckled when empty at 10,000 feet on return to sea level and a new tank was supplied but the filling pipe was altered in position and fouled the flap control wheel and chain, the starboard seat bearer and one of the fuselage bracing wires.[40] It proved to be a IIIB and not a IIIC petrol tank.

In February 1919 it was reported that there were only six Fairey 3C seaplanes in existence, *but, providing that Eagle engines can be obtained, there should be no difficulty in producing more at Messrs. Fairey's works if an order is placed immediately.*[41] In order to convert IIIB to IIIC seaplanes the following increase in weight:

Increase weight of float modification	30 lbs
Wireless with reel, panels, etc.	150 lbs
Bomb gear, sight & frame, sliding panel & release.	76 lbs
Bombs	500 lbs
Photographic apparatus	20 lbs
Weight of original Type IIIC in Type Trial conditions including guns, ammunition & 120 gallons fuel.	4,800 lbs
	5,576 lbs

The safety of the IIIB modified to IIIC condition was then subject to an analysis. It was found that with a total load of 5,076 lbs the lowest factor of safety was 3½ assuming that Grade A spruce was used, which was understood to be the case. This *should be all right for straight-forward flying, although of course, it does not leave a large margin for anything in the way of fancy work.'*[42] In May it was stated that £25,000 had been made available to enable the conversion 25 Fairey IIIA or IIIB seaplanes to IIIC standard. 21 IIIB and 15 IIIA machines were in store at Renfrew. The author of this memorandum was *not quite clear whether it would be more economical to convert the former type or the latter; whereas new wings have to be provided for the Fairey 3 B's, floats have to be provided for the 3 A's, the wings of which are similar to Fairey 3 C wings.*[43] In January 1920 it was noted that the IIIC would probably become a standard type in the RAF, but it had never been properly stresses, *the IIIb type have been taken in error and the load increased for the Eagle engine.*

Alterations had to be made to the Rolls-Royce engines to fit the IIIC and this was to take place in one of Fairey's shops. Trestles were arranged and eight engines set up, however Rolls-Royce only sent two men *and though it does not fall on us to have anything to do with the alterations of these engines, we fear that the labour that has been put on now will be insufficient to supply us with the quantity of engines necessary for the erection of the machines.* The Company requires at least five machines per week and any delay would of course delay the delivery of the aircraft.[44] Because of the urgency for the IIICs for North Russian Expeditionary Force, the seaplanes were flown early in the morning from the Thames at Isleworth directly to Grain instead of being re-assembled and test-flown at Hamble.

It was ascertained that the Standard Oil Tank and Auxiliary Petrol Tank *would be unnecessary; in fact; they would be undesirable in view of the heavy load required to be carried. I have therefore informed Mr. Fairey direct in order to save time…I regret that the correct procedure was not followed in this case owing to the urgency of the matter*. (author's underlining).[45] It therefore appears that 32 IIIBs were now being converted to IIICs. The tanks were to be omitted from the 5th machine making 28 in all to be so modified.

In May HMS *Pegasus* reported on problems with the radiators of the IIIC. The Eagle VIII No. 458AA installed in 5250[46] suffered damage to all 12 pistons, cylinders and valves due to faulty soldering in the port radiator. A seam gave way leaking in five separate places along its length and although the pilot landed immediately on seeing the temperature beginning to rise, the loss of water was so rapid that all cylinders and pistons suffered accordingly.[47]

The exhaust pipe was also the subject of criticism as a Memo to the Director of Research noted correspondence from the CO of the RAF unit operating from *Pegasus* in North Russia and requested *if it is possible to at once take steps to modify the design of both radiators and exhaust manifolds in the Fairey 3C machines still to be produced.*

If this is possible I should like to put up a requisition to replace all the radiators and exhaust manifolds in the machines now in service.[48]

There had been no previous reports of such problems with the radiators and it was felt that the radiators were all of sound design and should stand up to the work required if a new flexible mounting was fitted. This had been designed by Fairey and approved. This, it was considered, would obviate any further trouble.

The exhaust manifold defect appears to have been related to the manifold breaking. It was noted that the manifolds in use in other waters had no problems and it was requested that the broken one be returned so that an examination could be made as to its construction.

Production IIIC seaplanes were delivered to Nos. 229 and 230 Squadrons in November. The first IIIC was delivered to Great Yarmouth naval air station in November when the Germans capitulated, and had no chance to perform its role as a fighting seaplane, however the type did enter operational service in the North Russian Expeditionary Force in 1919. As discussed in the Fairey Campania chapter, the RAF had seaplanes in Russia from 1918. In 1919 the fighting Fairey seaplane was seen as being of use in the proposed campaign in Northern Russia and the introduction of the Fairey IIIC into service was pushed forward in the effort to equip the Allied force.

Russian Intervention

After the October Revolution, the Soviet Russian Government's signing the Treaty of Brest-Litovsk with Germany, freeing German troops for the Western Front, led the Allies to intervene in Russia. The reasons for this were many and complicated, as was the civil war that was fought between the Soviet Government and anti-Bolshevik forces. The port of Archangel in norther Russia lies on both banks of the Northern Dvina river near where it exits into the White Sea. Large quantities of supplies had been unloaded at Archangel by the Allies in order to keep Russia in the war, and on 2 August 1918, the Allies backed a coup against the local Soviet government and British naval vessels sailed into the port. Included amongst the warships was HMS *Nairana* with its complement of seaplanes. The Allied powers occupied Archangel, the British Gen Poole declaring martial law and virtually ran the city. These early arrivals were supplemented by Fairey IIIC floatplanes arriving in late April 1919 on board HMS *Pegasus.*[49] Short seaplanes did not arrive until the *Argus* came out in August.

A review of the RAF operations in North Russia found that aeroplanes were not considered suitable in North Russia due to: -

(a). Lack of aerodromes.
(b). Sites found near the railway were limited in size.
(c). The great amount of work required to render a site suitable for an aerodrome – drainage and planking.
(e). The nature of the country making a force landing either in the forest or the lakes.
(f). The nature of available aerodromes meant that only skilful and experienced pilots could be employed. This was illustrated as the only pilot lost from the RAF contingent was flying an aeroplane.[50]

Hence the amount of work that the seaplanes undertook during the intervention.

In May 1919, it was thought that *once the Devina* (River) *is free of ice it is to be expected that the enemy flotilla, that has been stationed all the Winter at KOTLAS, will advance down-stream against out positions in front of KURGOMIN and SHUSSUGA.* To meet this threat *every preparation has been made to insure successful collaboration between the R.A.F. and our Artillery*

Above: N9244 with Fairey and Short seaplanes. Note the white band around the fuselage.

Above: An example of the type of bridges that were attacked. Incendiary bombs were cabled for to render damage more lasting.

Above: A typical wood burning train on a wooden bridge. Wood was in plentiful supply from the local saw mill.

Above: N9239 at RAF Syren, Medvezh'ya Gora, Summer 1919. PoW working in front. The railway line is being carried to the beach to lay a plank runway. Guy Blampied recorded that the anti-Bolshevik PoWs wanted to work for the British and the especially good ones were trusted with all kinds of jobs. A Russian PoW caused some consternation one day when an inspecting Army officer found him loading Lewis ammunition trays.

Above: 2/Lt B.G. Blampied in June 1919, after bathing in the lake. N9235 in background. (G.B. Blampied via P Tomaselli.)

Above: Lt Harvey was the observer in N9235 when she was shot down. (G.B. Blampied via P Tomaselli.)

Above & Above Right: The remains of N9235 after the salvage party had removed the trees from the wreckage. (G.B. Blampied via P Tomaselli.)

Above: N9235 appears to be in the process of being refueled by using 4-gallon cans of petrol. Obtaining non-contaminated fuel was also problem in Russia. The object on the fuselage side near the cockade appears to be a camera housing. Reconnaissance was one of the Fairey's main duties.

Above: N9238 being hoisted out from HMS *Nairana*.

Above & Below: Two photographs taken from HMS *Pegasus*, showing the unloading of N9252, Divina River.

Above: Rolls-Royce Eagle powered Fairey IIID N9451, from the post-war batch N9450-N9499, on 25 July 1924. The early IIID seaplanes were very similar to the IIIC. The majority of the IIID series had the Napier Lion engine.

Concerns about the strength of the Fairey IIIC led to a machine being load tested on 24 November 1920. The aircraft was inverted and supported under the vertical struts in the front portion of the fuselage on beams arranged to give a single point of support. The fuselage vertical struts were strengthened by means of auxiliary struts. The fabric was stripped from the main planes and boards were laid across the front and rear spars to take the loads. At a load factor of 5.27 the internal anti-drag wire in the second bay from the inner end of the bottom starboard plane failed, by the ends pulling out of the wire wrapping. This allowed the front spar to buckle laterally. (TNA AVIA6/5897)

whether Naval or Military.[51] The Northern campaign was fought along the Murmansk to Petrograd railway line that went through the thickly forested areas. Both sides relied on the railway for transportation and supplies. Armoured trains were used that carried their own artillery pieces. As well as the hostile terrain, the areas of the fighting were close to the Arctic Circle and the winter nights were long and cold, while in the summer there was 24 hours of daylight. It was proposed to capture Medvezh'ya Gora giving access to Lake Onega. It was from this lake that the Navy Flotilla and the RAF seaplanes would operate. The banks of the river were surrounded by inaccessible snow forest. The use of landplanes was considered but abandoned due to the distance the nearest suitable site was from the front. Gerald Edward Livock, DFC, AFC, was a volunteer with the RAF for the North Russian Expeditionary Force in 1919. He was to be stationed on Lake Onega, in Eastern Karelia, that he described as being more of an inland sea rather than a lake, as it was over 100 miles long from north to south and from 20 to 30 miles wide. The sandy shore of the lake was ideal for floatplane operations.[52] The site selected proved to be ideal. *A railway siding ran through the Base, thereby obviating transport difficulties. The water was seldom rough enough to hinder the work of machines. The country being liberally intersected with lakes, the danger due to engine failure from flying seaplanes over land were consequently reduced. The type of engine supplied with the FAIREY IIIC was the most reliable in existence.*[53]

The RAF were based at the extreme north of the lake, where it fairly narrow. The unit was called the Syren Force.

The Allied Flotilla on the lake comprised a flagship, the *Jolly Roger,* an old triple-screw submarine chaser armed with a 3 pounder, and a 37 mm gun that had been stolen from a French armoured train. A number of old submarine chasers and small motor-boats armed with a variety of machine guns and/or 37 mm guns, made up the rest of the Flotilla. The aircraft were delivered by the *Nairana, Pegasus* and *Argus* to Murmansk as detailed in the following table. *Nairana* went up the Kola inlet and anchored off Murmansk on 28 May. *Pegasus* was already there. On 2 June the Fairey seaplanes were unloaded onto trains, the first train leaving at midnight.

The seaplanes were then transported by rail to the lake. The RAF unit had a flight of landplanes, R.E.8 and Sopwith Camel biplanes, and a flight of half a dozen Fairey IIIC floatplanes. The unit was operating in support of the remnants of the White Russians fighting against the Bolsheviks.

The Northern Dvina flotilla of the Soviet Army included several destroyers, and the Flotilla would be no match for them without air support. The attacks by the British damaged most vessels of the Soviet flotilla causing them to retreat up the river or risk being sunk. The Grigorovich M-9 and M-20 flying boats of the Northern Air Battalion appeared only rarely and when challenged, withdrew under the cover of their own anti-aircraft batteries. Enemy air raids took place by single machines on:

19 July – two bombs dropped off the bow of *Borodina.*
27 July – one bomb near *Hyderabad.*
28 July – Two bombs, one hit the Ordnance Depot barge. But did little damage

In addition to reconnaissance and bombing, the British seaplanes directed the flotilla's and ground batteries fire.

The first seaplanes arrived at Lake Onega by rail on 6 June. The seaplanes were erected that day and the seaplanes were met by a railway breakdown crane that was used to launch the seaplanes into the water. N9235 made the first flight with Lt Young with Lt Ross-Smith as his observer. The front line was about five miles away. The reconnaissance flight returned with a bullet through the airscrew. The anti-aircraft fire was to prove surprisingly accurate as events unfolded.

On 7 June, N9241, piloted by Lt Young with 2/Lt Eddes, left at 1400 to escort a small motor boat proceeding to Shunga. Then to take oblique photographs of the Shunga and reconnoitre the coast to Zagori and to patrol against hostile gunboats. Sighting *3 small torpedo boats and very large motor launch (bigger than any craft in Moduya Gora) lying in a channel between TCHALMUTSKAYA and Zagorie and proceeded towards them. On our appearance A.A. fire was experienced. Bursts were white and from a three pounder or larger gun. Attacked with one 12 lbs bomb which fell close to one boat. Boats were handled by experienced crews for a rapid fire was maintained from them and they maintained formation well, As it became apparent that we were to bomb them, they zig-zagged and maintained formation turning o as to get away to the southward, speed about 10 knots (cannot estimate accurately). Attacked with Lewis Gun, emptying three pans into them. Boats had by this time retreated to South of Island. Photographed these boats when bomb dropped and once afterwards. They were all flying the Red Flag. At this point the engine started popping and we turned for home, ran for ten minutes then gave out altogether. A landing was effected by gliding at a point 4 miles N. of Shunga. Shipping observed in addition to the three small torpedo boats. Only one vessel was observed in Zagorie Bay, about the size of a river passenger craft in the Thames or a brig about 60 ft long.* N9241 taxied home arriving back at 1445.[54]

The Red flotilla put in its expected appearance on the night of 8/9 June. Spotted by N9241 they were attacked with bombs and machine gun fire. When their Lewis gun jammed, Haines and Eades returned to base. N9236 and N9235 left to attack the enemy and were soon joined by the refuelled and rearmed N9241. The flotilla and the seaplanes attacked the Soviet vessels. N9235 with crew of Blampied and Macnab, left at 0200 to attack *Dropped 7 20 lb bombs all in close proximity. 1 V.C* (Very Close?) *straddle obtained. Returned to base* (at 0305) *owing to MG trouble which could not be rectified.*[55]

The three seaplanes attacked one after the other with bombs and machine gun fire, straddling one steamer with two 20-lb bombs. Concentrated machine gun fire on the ship's gunners and bridge of the largest ship caused the vessels to turn and retreat. The anti-aircraft fire was heavy and accurate, one hit on

N9236 missing the observer by three inches. General Maynard was to write that the *bold attack of our little craft might not have met with success had they been operating alone…Fortunately for them, they could now rely on seaplane assistance.*[56]

Livock recorded a different opinion of the anti-aircraft capability of the enemy. He made his first operational flight in Russia on 6 August, and would make a flight of approximately two hours every day. These flights were made in co-operation with land forces and comprised shooting up enemy strong points, bombing railway stations and villages, searching the lake for enemy maritime traffic and making a general nuisance of themselves. According to Livock, there was no air opposition and the enemy had no effective anti-aircraft guns, the RAF could fly as low as 3,000 feet for accurate bombing.

Attacks were carried out on the trains and train line. On 11 June, Isaac and Eades, in N9236, blew up the line. They came in at tree-top height and dropped a 230-lb bomb with a delayed action fuse. This tactic was to be adopted and used again.

There were few maps of the area and these were often wildly inaccurate. In order to take vertical photographs, a hole was cut in the bottom of the fuselage. This had to be covered up when taking-off or else water was taken in and ran down to the tail making taking-off impossible.

The railway bridge at Fedatova was successfully bombed by N9328 on the 21st. Unfortunately, an engine failure meant that a forced alighting had to be made on the lake. N9237, piloted by 2Lt B.G. *Guy* Blampied with Lt Smith, sighted the drifting seaplane on their return from a bombing raid on Dianovi Gori, and alighted near the stricken Fairey. A line was passed to N9328 and it was towed for half an hour until the *Atlanta,* an American launch, came across the pair and took the seaplane back to base.

On the 29th Haines and Eades followed a train until it reached a straight section of track south of Kapaselga, and dropped a 2½ second delayed bomb that landed on the track and the train ran over it as it exploded. The fuel wagon was demolished and the rear of the engine ended up in the bomb crater.

It was while flying a reconnaissance the next day to confirm this report that *Guy* Blampied, with Lt Harvey as observer, was shot down in N9235. Guy wrote that the *Bolshy ground fire was poor their A.A. guns never bothered us at all. When I was shot down, it was my own fault. I was turned out of bed about 1 a.m. to do something I cannot remember what but the wind was blowing on shore so I took straight off & did not gain sufficient height before flying over enemy lines.*[57] The result was that he was hit *in petrol tank and radiator, had to land in trees. We were in Bolshy lines but after swimming a river and six hours walking through the woods came to Lobskagora, where we got a peasant to take us to MG* (Medvezh'ya Gora). *Arrived at 12 o'clock.*[58] They drew a picture of a horse and cart for the farmer which he produced and drove them back to base where he was rewarded with plenty of rations. *Guy* was to write later that *I did not realise that radiator was hit and lost water till I saw exhaust pipe was red hot which with petrol floating about from tank that was holed was not a happy thought so I turned off petrol supply to engine & hoped I could reach the lake but the engine ran out of fuel before I could get to the water so had to land on trees.*[59]

The pair were extremely lucky to survive their force landing in the trees. Photo reconnaissance showed that a new track had already been laid around the wreck. A salvage team of five airmen under Lt Noonan proceeded five miles along the south bank of the Unitsa River then one mile inland to return to the wreck to get spares but did not stay long due to the proximity of the "Bolshys".[60]

The attack by Haines and Eades caused the Soviets to stop running their trains in daylight and only run their trains at night. This prevented the rapid movement of troops and heavy artillery. On 1 July, Lettington and Radfdord attacked a paddle steamer scoring a direct hit causing the boiler to explode and the crew to abandon their vessel. On the 8th, the Allied Flotilla suffered a loss when the *Jolly Roger* blew up killing five of her crew. Luckily two Fairey were able to alight and pick up survivors. The arrival of submarine chasers from Britain overcame the loss of this the largest Allied vessel.

The British Government made a decision to withdraw from Murmansk and Archangel before the coming of winter. A final offensive was planned to create the necessary breathing space for White Russian troops to move up to the line. Reinforcements were brought out by the *Argus* and Sopwith Camels and R.E.8 landplanes as well as seaplanes, and included amongst the new aircrews was Capt G.M. Livock.

On 3 August a combine operation saw four Fairey seaplanes cover an amphibious attack on Tolvoya. This was coordinated with White Russians who had been advancing down the Shunga peninsula. The Soviet flotilla was caught in the harbour and two vessels ran aground, their crews abandoning them and escaping into the forest. Two more were pursued until they surrendered to the Allied vessels. In an hour the main Soviet naval presence in the area had been destroyed. The captured vessels were pressed into Allied service.

The main attack on the Dvina front took place on Sunday 10 August at 10 am. The monitors *Humber, M.27, M.31* and *M.33* opened fire at 11.20 a.m. Spotting was carried out by seaplanes, kite balloons and Naval Observation Station in advance positions on shore. Weather was so bad it was very difficult for the seaplanes to operate. *Cicala* was therefore brought up to counter the enemy flotilla with her 6-inch longer range guns. She came up slowly due to having earlier suffered damage from a mine. Soon after noon the enemy was reported retreating up river.

On 14 August 1919, a force of Russian infantry and British machine gunners advanced up the left bank road and took Puchega with little resistance. That same afternoon s party went by seaplane to Puchega, *boarded the Enemy Gunboat "Scorpion", sunk by bombing and field gun fire just above the village, bringing back the breech-block and dial sight of her 75. m/m gun in the seaplane.*

The seaplanes did exceptionally fine work, flying in thick, wet

Fairey IIIC in Russia

Serial	Ship	Unit	Date	Notes
N2233				In Soviet service.
N2236	*Ark Royal*		by 26.09.21	South Russia.
N2239	*Pegasus*		by 10.20	South Russia. Returned Malta 11.20.
N2241	*Ark Royal*		by 21.07.21	South Russia. Returned.
N2243	*Ark Royal*		by 14.02.22.	South Russia.
N2247				In Soviet service.
N2249	*Ark Royal*		by 04.11.21.	South Russia.
N2250	*Ark Royal*		by 29.06.21.	South Russia.
N2253	*Ark Royal*		by 09.06.21.	South Russia.
N9230	*Nairana*	Syren Force	18.05.19	Used by Soviets after capture from White Russians.
N9231	*Nairana*	"	by 28.08.19	Used by Soviets after capture from White Russians.
N9232	*Nairana*	"		Not left when withdrawal of British forces.
N9233	*Nairana*	"	by 08.06.19	Crashed 31.08.19.
N9234	*Nairana*	"	by 11.08.19	
N9235	*Nairana*	"	by 18.05.19	First flight from Lake Onega 06.06.19. Brought down by AA fire 30.06.19. Complete wreck. by 09.07.19.
N9236	*Nairana*	"	by 18.05.19	Forced landing 05.08.19.
N9237	*Nairana*	"	by 18.05.19	
N9238	*Nairana*	"	by 18.05.19	Force landing 21.06.19. Towed to base.
N9239	*Nairana*	"	by 18.05.19	Flown 11.06.19 only.
N9240	*Nairana*	"	by 18.05.19	
N9241	*Nairana*	"	by 18.05.19	Used by Soviets after capture from White Russians. Crash and written off 22.01.1922.
N9242	*Pegasus*	Dvina River	by 27.04.19	Used by Soviets after capture from White Russians.
N9243	*Pegasus*	Dvina River	by 27.04.19	Nairana. Syren Force by 14.09.19.
N9244	*Pegasus*	Dvina River	by 27.04.19	
N9246	*Pegasus*		by 24.01.21	South Russia.
N9247	*Pegasus*	Dvina River	by 27.04.19	
N9248	*Pegasus*	Dvina River	by 27.04.19	
N9249	*Pegasus*	Dvina River	by 27.04.19	Shot down 14.07.19. Crew of 2/Lt H.L. Marshall, Lt G Lansdowne, PoWs. Into Soviet service.
	Ark Royal		17.06.21.	Not left in Russia.
N9251	Pegasus	Dvina River	27.04.19	20.07.19. A bomb exploded on take-off. 2/Lt A.J. Rankin was wounded but his observer, 2/Lt J Gondré was drowned.
N9252	Pegasus	Dvina River	27.04.19	
N9253	Pegasus	Dvina River	27.04.19	
N9254	Pegasus	Dvina River	27.04.19	
N9255	Pegasus	Dvina River	27.04.19	

weather, bombing enemy ships, etc., and carrying out valuable reconnaissance work. Two direct hits, one which put "Scorpion" out of action, were recorded, and they contributed in very great measure to counter the fire of the Enemy Gunboats when their support could have been of the utmost value to his Shore Forces.

The kite balloon did good work when weather permitted. *It proved a most effective guide to stragglers, and the colonel Commanding the Left bank Attack, who got separated and was missing for two days, working his way back by it. At night it was lit up by searchlight from time to time.*[61]

On 18 August N9233, N9234 and N9238, conducted a raid on Petrozavodsk, and important port about 80 miles behind the lines. The three floatplanes carried out the raid successfully but Livock had to nurse a "dud" engine on the way back. Livock recalled leading another raid by three Faireys, an R.E.8 and four Camels to bomb the Suna Bridge. On the return journey he noticed that Sykes, one of the Camel pilots, was missing. Sykes was forced down by engine failure and commenced to walk out but was captured by locals who handed him over to the Bolshevik authorities. He later said that he was well treated in their hands. When the Flight caused an air raid warning to be sounded, he was marched to the docks at Petrozavodsk and left there while his guards hid. He considered himself safe as he was standing on the Flight's target! He was to return safely to England.

Operations from Popoff

Popoff was the port for Kem in the White Sea. In August a unit was required at short notice to watch the road from Onega to Scroka. In order to maintain the reconnaissance flights from Popoff Quay while the *Nairana* was at Archangel, 50-ton floating sheerlegs were utilised. Considerable skill was required in handling machines on the water in the absence of the ship. The seaplanes would come to the sheerlegs under their own power and were then secured, a difficult job, especially if the wind was blowing towards the quay. Daily reconnaissances were maintained without any machines being damaged. The distance to Onega *in bad weather taxed the Fairey 3C to the full – machines frequently returning with only petrol for a further quarter of an hour's flight.*

When carried in *Nairana,* this unit spotted for bombardment of Onega. The CO of *Erebus* acknowledged that the spotting was as good as any that the ship had obtained during two years firing on the Belgium coast. It appears that the *Nairana* carried two seaplanes during this operation.[62]

On 6 August 1919, the White Russians had established the Murmansk Aviation Division. It had a landplane and a seaplane flight. The pilots trained on Short 184 and Fairey IIIC floatplanes. The seaplane flight had two Fairey, three Short seaplanes and three Grigorovich flying boats.[63] By 10 September the British flotilla on the Northern Dvina River had retreated up to Troitsa, and then gradually started to move downstream towards Archangel. Some serviceable seaplanes were returned to the UK, but several were left with the White Russians who did not counter the Soviet Army for long, and at least five Fairey and three Short 184 seaplanes were captured by the Soviet Army.

The first Fairey IIIC that the Soviets acquired was N9249 that Lt G Lansdowne and 2Lt H.L. Marshall were flying on 14 July 1919, when they were hit by fire from Soviet gunboats. The fuel tanks were punctured but the crew did not notice until they ran out of fuel and were forced to alight on a river and were taken PoW. The Soviet tugboat *Trud* picked up the seaplane and took it to a naval base where it was dismantled and sent to Petrograd. It entered service with the 1st Hydroaviation Battalion in late October.

On 27 August 1920, a Commission accepted ex-British material arriving from the railway station of Segezha. The following Fairey III aircraft were then stored at the Gutuevsky Island naval aviation base in Petrograd:

N2233 (c/n F208). Fitted with one Vickers gun, without wings, crated in good condition, fit to become operational.
N2247 (c/n F294). Fitted with one Vickers machine gun, lacking underwing floats.
N9230 (c/n F307). Without machine gun, magneto and compass.
N9231 (c/n F308). Without machine gun, magneto and compass.
N9241 (c/n F318). Without machine gun and with badly damaged underwing floats and main pontoons, fit to be used for spare parts.
No number. Damaged from fire, crated without airscrew, fittings and undercarriage, fit for use for spare parts.
No number. Badly damaged fuselage, fit to be used for spare parts.

Five Short 184 floatplanes in good condition were also received. The aircraft required attention and thus were fielded only gradually.[64]

On 4 September 1920, the fuselage of Fairey III N9233 were turned over to naval aviators from the Komendantsky airfield storehouse in North Petrograd.

The listing of "Western Front Hydroaviation Strength" of 15 November 1920, gave Fairey (N)9242 with the 1st Hydroaviation Battalion, and (N)9230 with the 1st Hydroaviation Unit.

Three Fairey and one Short seaplane took part in the operations against the Kronstast insurgents and sailors of the battleships held in the ice. On 17 March 1921 (new style), N9230 piloted by M.C. Lindel with mechanic Ignatiev as observer, dropped four 35-lb bombs on two of the ships and saw a large white cloud appear on one ship that indicated that they had scored a hit and burst a steam pipe.

On 15 September 1921, the 3rd and 4th Flying Boat Detachment used one and two Fairey seaplanes respectively. That autumn there was combat with Finnish nationalists on the

Karelian border. The 1st Reconnaissance Hydroaviation Unit with two Fairey and one Short 184, deployed to Lake Ladoga, one of the Faireys crashing on 22 January 1922.

The Fairey floatplanes operated on the Baltic Sea until the spring of 1923. They were replaced by Italian Savoia 16bis flying boats.

Post-Armistice

As at 30 November 1918, there were in commission:
50 Fairey IIIA
20 Fairey IIIB with 34 still on order.
1 Fairey IIIC.

For comparison there were in commission:
22 Short 184 (Early models)
294 Short 184 Maori.[65]

With the end of the war the Fairey Aviation Co was in need of orders to keep itself intact as a working aviation enterprise. To this end N10 was purchased back from the Government. A Fairey advertisement from *The Aeroplane* 20 October 1920, praised the IIIC under the title "A Maker of History" in the following words:

She was delivered to the R.N.A.S. in April 1917; She was on Service till after the Armistice; She was bought back in May, 1919, as a "Disposal" Machine; She was used to carry newspapers from Blackfriars to Thaner; She was used for Experimental Work during the Summer of 1919; She was flown in the Schneider Cup Race in September, 1919, and was the only competitor to return intact under her own power; She was used with experimental Amphibian Undercarriages during the early part of 1920; She did Ferry Work between Southampton and Sheerness; She has survived Nine Different Engines.

In this work no Single Component Part, except the Undercarriage, has been replaces. Fuselage, Wings, Tail-plane, Elevators, Fins, Tail-float, even the Engine-Bearers, are those originally fitted. The Wings have only been re-covered once, the Tail-Unit still has the original fabric. The Cock-pit has been altered to carry one, two, and three people.

Despite all this Her Performance Equals that of the Best Modern Flying-Boats.

Anther advertisement with the same "History" but a photograph of the N10 with land undercarriage appeared in the following week's journal and continued until 10 November. C.G. Grey using the wording of the advertisement in his editorial of that issue.

The N9 and N10 had been bought back from the Government. N10 was given the civil registration G-EALQ. Its first public appearance was at Bournemouth for the Schneider trophy race.

Schneider Cup

The Fairey entry for the 1919 Schneider Cup race was the N10 fitted with a twelve-cylinder three-bank liquid-cooled 450-hp Napier Lion engine and wings of reduced span. According to *The Aeroplane*, Mr Fairey did not expect to score particularly on speed although it would require something very fast to beat him. The Fairey chassis was thought to give the machine an advantage if the weather was rough as it would be able to get off and land while a pure racing seaplane would inevitably crash.[66] The modified N10 was flown by Lt Col Vincent Nicholl, DSO.[67] In addition to having a single bay due to the clipped wings of 28 feet span, the standard nose radiator had been replaced by one on each side of the fuselage. Starting was by ballot with Nicholls first one off. The Fairey entry withdrew from the race due to the fog, Nicholl nearly hitting a sailing boat when flying at 15 feet looking for the mark-boat. He alighted and beached his machine. He explained to Mr Fairey that if he could not see the mark-boat it could not see him. The gamble for rough sea conditions had not paid off, conditions on 10 September were calm. Despite this, the race was a shamble with the Supermarine entry sinking due to a holed hull, the Sopwith entry being beached in a hurry due to the floats leaking and promptly wrecking them. The Italian entry made the most circuits but missed the mark-boat and his time declared void. However, the Fairey was the only competing machine that flew home.

Following the race Richard Fairey wrote to the Air Ministry that he had fitted a Napier Lion engine in *a standard Type III machine for the purpose of the Schneider Cup Seaplane Race. The machine is in fact the original Type III. No. N.10 which we purchased from the Government when over 2½ years old.* Fairey wanted the machine to have an official type trial *in order to place the performance on record both for comparison with the standard Fairey III.C. fitted with the Eagle VII. Roll Engine and other types.* Fairey was willing to pay for the tests and while he would like to have the test done at Hamble, he would transport the machine to Grain if this was more convenient.[68] This was agreed to and the trials were to be at Grain at Fairey's risk. *Provided the conditions are accepted by you, no charge will be made in respect of the trial.*

The Air Ministry organised a competition for civilian aircraft with an entry for amphibians. G-EALQ was entered with a special combined float and retractable wheel undercarriage. It reverted to its original two-bay wings and had a two-seat side-by-side cockpit behind the pilot. The twin floats were long with nine watertight compartments. The tail float was only for when the wings were folded, otherwise the single-step floats supported the whole aircraft. The Fairey entry came third. G-EALQ was operated from Hamble as a communication hack until finally retired in 1922.

At least four Fairey III seaplanes were to receive civil registrations. Maj W.T. Blake, with his cameraman Geoffrey H Mailins, had secured sponsorship of the Daily News to attempt to fly around the world. The pilot was Capt Norman Macmillian. The plan was to use four different aircraft for the different stages of the proposed journey. A D.H.9 for the first stage to Calcutta; a Fairey IIIC from there to Vancouver; another D H.9 to Newfoundland, and finally, a Felixstowe F.3 flying boat for the final return to the UK.

G-EBDI (ex-N9253) was shipped out to India and erected by the RAF, and was awaiting the adventurers when they arrived after nearly three months and many adventures. Blake suffered from appendicitis and had to enter hospital Macmillian and Malins then took the IIIC and proceeded to fly the next section. Leaving on 19 August 1922, they were forced down with engine failure. Macmillian made a successful alighting on a rough sea. Macmillian managed to get the engine to start and they taxied over the rough waves to the island of Lakhidis where they ran aground on a mud bank.

They were to stay there for two days and three nights before conditions allowed them to attempt to take-off. There was only an hour when the tide allowed for a successful attempt. Leaving on the 22nd they were only 15 minutes into the flight when the engine again began to misfire. They alighted and started to work on the engine. The starboard float had been found to be leaking and had been pumped out before take-off, but now it was realised that by the time they had repaired the engine it would have taken in too much water for a successful take-of, so they started to taxi again. The machine finally settled into the water, as the fuel gave out when in sight of land. They climbed out onto the port wing, discarding everything but essential records. The machine finally heeled over backwards and floated upside down mainly due to the empty fuel tanks providing buoyancy. They spent another day and night on the wreck until the upper wing became stuck on a mudbank. They were finally picked up by a launch that had set out in response to a message that they had had sent from Lakhidia. Efforts to salvage the aircraft ended when the tow rope broke and the machine sank. The flight was then abandoned.

Lt-Col G.L.P. Henderson purchased IIIA G-EADZ (ex-N2876)[69] that he had converted to IIIC standard and reregistered as G-EAMY. According to file notes, no CofA could be made for this machine as drawings and details of modifications made were not forwarded for approval. Also, since preliminary inspection by the Department, the machine had crashed, been repaired and modified with the floats being replaced by a land undercarriage without notification – this being a contravention of the Air Navigation Acts. In February 1920, the application for a CofA for this machine was cancelled and the fee returned to the owner. Henderson also bought IIIC N2255 that was registered as G-EAPV. No official application for a CofA had been received for this second machine.[70] These two seaplanes were converted to have two passenger seats immediately above the rear fuel tank and two seats in the rear cockpit.[71] *The Aeroplane* recorded that three large Fairey seaplanes left the UK on 23 January and had landed near Copenhagen, one being damaged when it broke its airscrew on landing. They were flown by Lt Col G.L.P. Henderson, Maj Farrar and Lt Jewell, en-route to Sweden. The machines owner was identified as Henderson.[72] Further editions of *The Aeroplane* gave details of Henderson's activities. The machines were operated by P.O. Flygkompani that had the *three Faireys, one A. W., three Avros, and three D.H.6.* [73] G-EAMY, another of the Faireys, could be fitted with wheels, floats or skis for landing on land, water, ice or snow, and presumably the other two could also be converted as necessary. In early March Henderson with Lt Youell flew one of the Faireys to Are and on arrival found bad visibility and could not judge the height of the machine above the snow and turned over on landing, fortunately without any injury to the crew, but the machine was severely damaged.[74] The final IIIC lasted until the end of the year when it was taken out of service. G-EAMY had crashed in August.

G-EAPV was used on experimental flights from the end of August 1920 to the construction site for hydroelectric dams at Suorvajaure, about 100 kilometres from the Porjus power station. This work site soon became a tiny settlement of about 400 workers and could only be reached by horse drawn sleigh or, in summer, by boat, taking two days. This was cut down to an hour using an aircraft. The Fairey IIIC was used but proved to be inadequate in the severe climatic conditions. What was needed was a small, easily controlled, stable aircraft with an air-cooled engine. Avro 504L biplanes were to be used, three being employed from October 1921.[75]

G-EARS (ex-N9256) was purchased by Faireys and used as a seaplane demonstrator. Sold to the Aircraft Disposal Co, it was shipped to Canada where it was registered G-CYCF. In January 1919, the Fairey Aviation Company had sent plans of the Atalanta, Campania and Hamble Baby to the Government, apparently in response to a Canadian request for information about the types. Although they forwarded the plans as requested Fairey also included plans of the *Type 3, 3B and 3C* machines that *we shall be glad if you will also forward to the Canadian Government as being more modern than the Campania or Hamble Baby Machines.*[76]

Post-war the RAF required a fighting seaplane to replace the venerable Short 184. The Fairey III series was seen as a possible replacement. In additional to their development, that had come to almost a standstill with the Armistice, the Fairey machines were being used for experimentation. For example: - Fairey IIIB N2852 (Sunbeam Maori) was being used for comparative trials of mineral and castor oils, Grain reporting that engine was being reassembled for the W/E 01 January 1919. Bristol Fighter C4900 (Falcon III) was also being used for similar trials. Fairey IIIC N2246 was to be tried with balanced ailerons.[77] The choice of the correct airscrew was an important decision to get the best results out of the engine/airframe combination. N2246 was trialled with different airscrews at the Marine Aircraft Experimental Station, Isle of Grain in March 1919. A four-blade airscrew from an F.3 flying boat was tried and compared with the original airscrew that had been manufactured to D.R.G. 6194.

N2246 was converted to a landplane in June 1920 and used for deck landing experiments on HMA *Eagle*. With the end of the war the shortage of Rolls-Royce engines had ended and the

Above: The F.10 at East Cowes in Schneider Trophy guise with triple row Napier Lion engine. It has single bay wings and the forward cockpit has been faired over, the machine being piloted from the rear seat. The civil registration, G-EALQ, was painted on the top of the upper wing.

Above: Fairey photograph of G-EAPV, their civil conversion of N2255. This aircraft was lost in Sweden when operating from Lt-Col G.L.P. Henderson's enterprise. It crashed in unknown circumstances on 27 July 1920, while on a scheduled flight and was written off. Both occupants were uninjured.

Above: One of the Fairey IIIC landplane conversions in Sweden.

Above: One Fairey IIIC, N9256, was eventually received in Canada as part of the Imperial Gift and was given the registration G-CYCF.

Report N.M.242				
Speed at	Original Prop	Revs	Boat Prop	Revs
2,000 ft	96	1940	88	1830
4,000 ft	94½	1915	86	1825
5,000 ft	94	1915	83	1800

Climb to	In Min - Sec		Rate of Climb in ft/min		Indicated Air Speed		Revs	
	Orig Prop	Boat Prop	Orig Prop	Boat Prop	Orig Prop	Boat Prop	Orig Prop	Boat Prop
2000 ft	2.20	2.45	794	710	59	53	1700	1660
5,000 ft	6.40	7.45	604	604	59	50	1720	1640
8,000 ft	12.30	14.30	422	422	59	50	1720	1640

The machine had a load of 4.440 lbs for the trials, the total weight when fully loaded was 4,800 lbs.
Source: AIR1/1204/204/5/2616.

Eagle version of the Fairey III series would have been wanted rather than the IIIA version.[78]

Cape Flight

On 8 January 1919, the Fairey Aviation Co was told confidentially that four Type IIIC seaplanes would be allocated to the England to Cairo and the Cape flight.[79]

The following modifications were made to the four Cape seaplanes:

1. Floats specially strengthened.
2. Guns, ammunition boxes, etc., removed.
3. Larger radiator fitted.
4. Comfortable seats fitted.
5. Wireless sets were to be continuous wave capable of transmitting and receiving 300 miles, fitted with telescopic mast, and so arranged to be used when seaplane is disabled.
6. Recognition lights fitted.
7. Boxes, as large as possible, fitted for crew's gear and spares.
8. Larger windscreen fitted.
9. The engines were modified by Rolls-Royce to low compression engines.

If any further alterations or additions for Pilot's comfort are required, may it be stated please, as early as possible.

Fairey replied that the *Machines were designed for fighting purposes and to this end the protective cowling screen was cut down to the minimum in order to improve the pilot's view.* It was suggested that a larger wind screen would afford more comfort to the pilot. It was also mentioned that the Vickers gun and ammunition boxes could be removed and a more comfortable seta fitted for the passenger who had *the mere gunnery seat now fitted.*[80]

The numbers of the Rolls-Royce Eagle VIII engines to be fitted to the Cape Machines were 2256, 2218, 2312 and 4100. All had been converted by Rolls-Royce to Mark VII. Rolls-Royce asked that the latest type epicyclic gear and strengthened crankshaft be fitted to these engines. This would take approximately four to six weeks. This was agreed to on 12 May 1919.[81] The flight did not take place, the IIIC seaplanes being urgently required for the Russian expedition. The flight eventually took place in 1926 by Fairey IIID biplanes fitted with landplane or float undercarriages as required over the various sections of their flight.

The use of the IIIC seaplanes apart from their Russian adventure is poorly recorded. In the summer of 1921 Air Commodore C.R. Samson was sent out to command RAF Mediterranean. This comprised a HQ at Malta, a seaplane station at Calafrana, Malta, a seaplane station at Feneraki, Constantinople, and HMS *Pegasus*, a seaplane carrier. At Calafrana there were five F.2a and one F.3 flying boats and two or three Fairey IIIC floatplanes, while at Feneraki, there were four Fairey IIIC floatplanes with a further five aboard *Pegasus*. The last Fairey IIIC seaplanes in RAF service were withdrawn from service with No. 267 Squadron in Malta, in August 1923.[82]

The Type IIIB & IIIC Described

The Manual on the type states that the Fairey Types IIIB and IIIC *have been evolved from and still possess the main features of the original type III No. N.10.*

The original type III machine was fitted with a Sunbeam 260 h.p. engine, and the type letters signifies the following modifications: -

IIIA The same with land or skid chassis.

IIIIB The same with extended upper plane and sea chassis.
IIIC Standard wings and sea chassis, and "Eagle" Rolls-Royce engine.

The uppermain plane of the type IIIB has an overhang of about 9 feet, while the standard upper and lower main planes for the type IIIC both have the same span.

All machines are adapted to carry bomb gear, machine gun, etc., and the equipment is interchangeable throughout the series. All the essential components are interchangeable, so that a standard fuselage may take either type of wings, chassis or tail unit. For changing the type of engine, special adaptors are used, so that the standard engine plate bearers may accommodate either a Sunbeam, Rolls-Royce, Liberty, Napier Lion or Hispano Suiza engine, by fitting suitable wooden bearers in the engine plates. All fuselages are provided with fittings to take either the land or sea chassis.

To facilitate transport and storage, the main planes may be folded back along the fuselage, quick release fittings and hinge pins being fitted to the end of the main plane spars for this purpose.[83]

The fuselage was the usual wire braced wooden structure with steel tubes where required. The engine bay occupied the first three bays with the pilot's cockpit occupying the next bay followed by the auxiliary and emergency petrol fuel tanks in bay 5, followed by the gunner's cockpit. The oil tank was beside the engine with three pressure fuel tanks in the fuselage. Under the pilot was a 47 gallon tank; behind the pilot were a 30 and a 34 gallon tank. The fuselage was wire braced. At the pilot's cockpit the necessary strength is given to the fuselage by inserting a bent ash member that passed completely round the top of the cockpit and was bolted to the top longeron. The pilot's seat was supported by wooden bearers. The main petrol tank was placed under the seat bearers. In the IIIB there were two petrol tanks in this bay, the pilot's seat being accommodated by the special shape of the top tank.

In bay 5 the IIIC aircraft was fitted with the auxiliary and emergency fuel tanks, the emergency tank being above the auxiliary tank, each tank being supported by a wooden cradle and held in position by bracing wires. In the IIIB no tank was installed in this bay but the wireless gear was fitted in the back of the same bay so that the gear was immediately in front of the gunner. This was the only difference in the fuselages of the IIIB and the IIIC.

The gunner's cockpit naturally had no top cross bracing but had a 18 SWG flat strip of mild steel 5/8 inches wide across the side of the cockpit. A Scarff ring could be fitted. In the IIIC the wireless gear was fitted on the port side. On the starboard side a horizontal wooden member was carried across the side struts and attached to the two vertical members of the framework in order to carry the wireless generator and the supports for the bomb sight. Five double drums of ammunition were stored in an ammunition rack fitted along the longerons. A movable hook was attached to each of the bottom longerons. The gunner could operate this by means of a flexible cable to hold the wings in the folded position.

The pilot's normal entry and exit was over the fuselage between the interplane struts. In an emergency the pilot could lean backwards and push open the emergency door constructed in the centre of the rear portion of the pilot's "hooding." By exerting an upward movement, the door and fabric portions of the "hooding" were released and fell back on to bearers that were capable of bearing the weight of the pilot. The top of the door was secured in position by brass hinge pins in slots and a vertical movement of ¼ inch was sufficient to give complete release.

The same aerofoil section was used on each type of wing. On all main planes movable control surfaces hinged at the rear spar occupied the whole span. On the IIIC they comprised the ailerons and wing flap, and on the IIIB, wing flaps only. What are referred to as wing flaps in the manual were the Fairey patented camber gear. The wing ribs therefore terminated at the rear spar; the aerofoil shape being continued by the movable surface ribs. The trailing edge of the ailerons and wing flaps was a mild steel section.

The IIIC had ailerons on both top and bottom wings. The IIIB had ailerons on the upper wing only, the lower wings having the wing flap throughout their span. The wing flaps were lowered for take-off and landing and for maximum speed in the air, they were raised. These flaps could be lowered to an angle of 45° when the wings were to be folded. Also, for folding purposes the upper planes had their training edge cut away at an angle of 45° at their root with a similar cut away on the upper centre section.

The wings were built up around two spars 2 feet 8 inches apart. The spars were solid and of I section. Compression members were built up of two grooved and tongued sections that joined to form a hollow member. Former ribs were spaced between the ordinary ribs and were attached to the spruce leading edge. At the wing tip the leading edge was an ½ inch mild steel tube which was carried around the wing tip.

Provision to anchor the wings to the fuselage when folded was made on the lower planes, the fitting being attached to the bottom of the rear inner interplane strut. When the wings were folded a light jury-strut was fitted between the ends of the upper and lower front spars. A wing-walk 8½ inches wide was fitted across the spars at the root of the lower planes.

Floats were positioned under the outer interplane struts. These floats were held in place by three streamlined tube struts and cross bracing wires. The bottom surface of the floats was inclined so that it became parallel to the surface of the water when the machine rolled. When the land chassis is fitted the floats were replaced by wing skids of the ordinary bent ash type.

The wings of the IIIB were braced by cables while streamline wires were used on the IIIC. The centre section struts were of steel tube.

Both types of chassis, sea and land, were composed of steel struts.

The single step floats were made of three-ply spruce held together by copper rivets, bolts and special structure joints. The floats were broken up into compartments by ply bulkheads. Inspection covers were fitted to the top of the floats with air

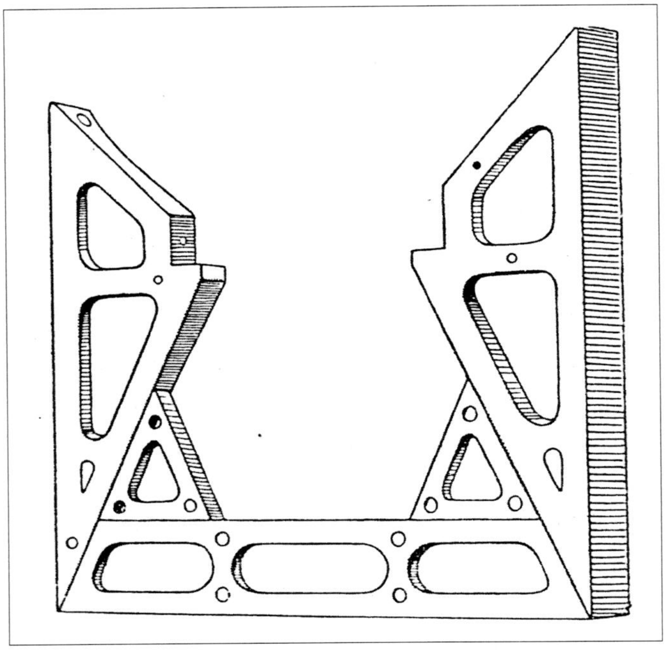

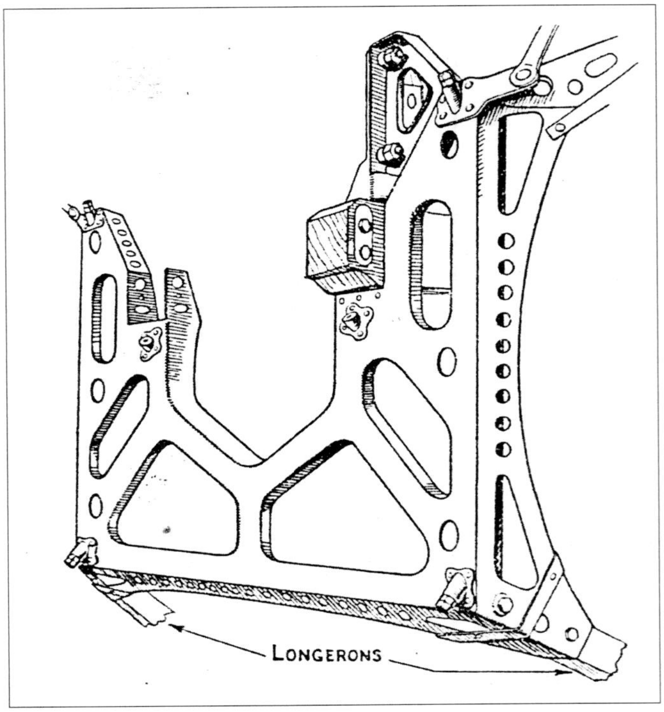

Above Left, Above, & Left: The side struts for the fuselage numbered from 1 to 14, the engine bay was between struts Nos. 1 and 4. Across side struts Nos. 1 to 3 were three steel Engine plates to support the engine bearers from the front to rear are shown.

Below Left: The main plane fittings for the hinged joint on the front spar are shown in these two views.

Below & Facing Page, Upper Left: The main plane folding fittings for the hinged joint on the rear spar area shown in these two views.

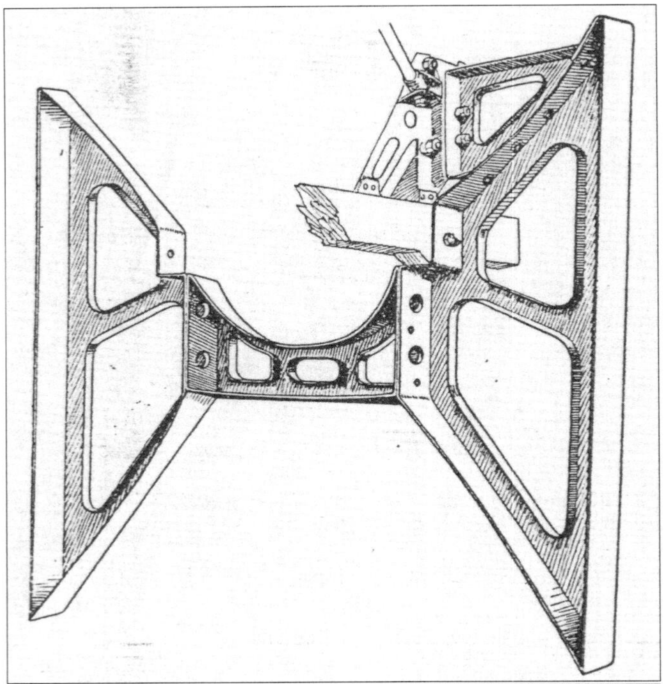

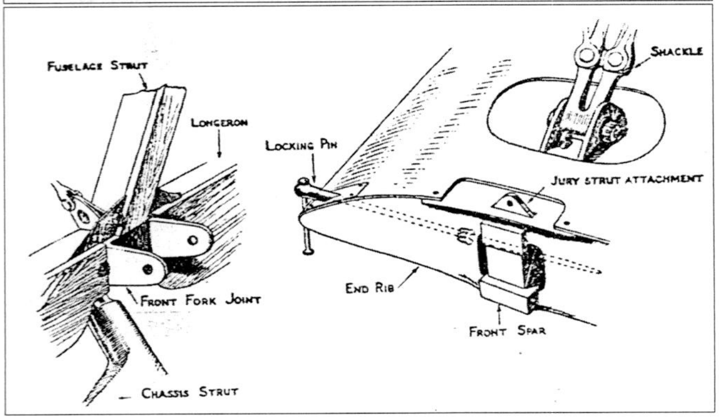

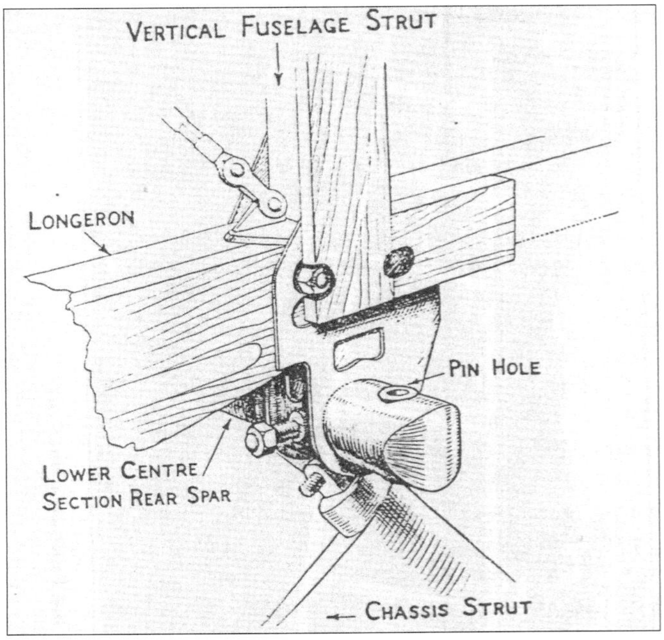

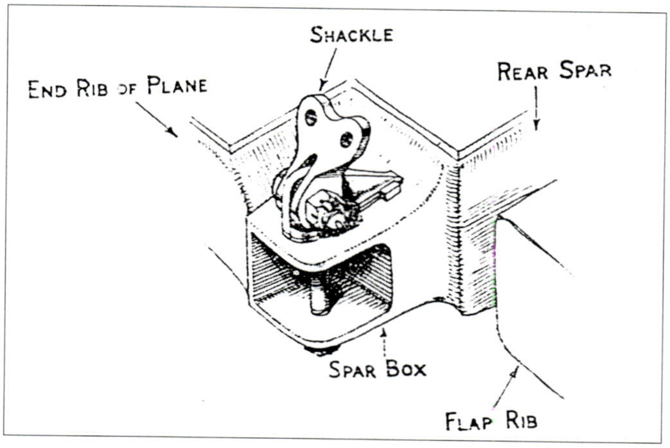

Right: Wing tip float. The floats were made of spruce rails and bulkheads with canvas sides. Note the shape to take spray away from the seaplane. These were replaced with wooden skids when used as a landplane.

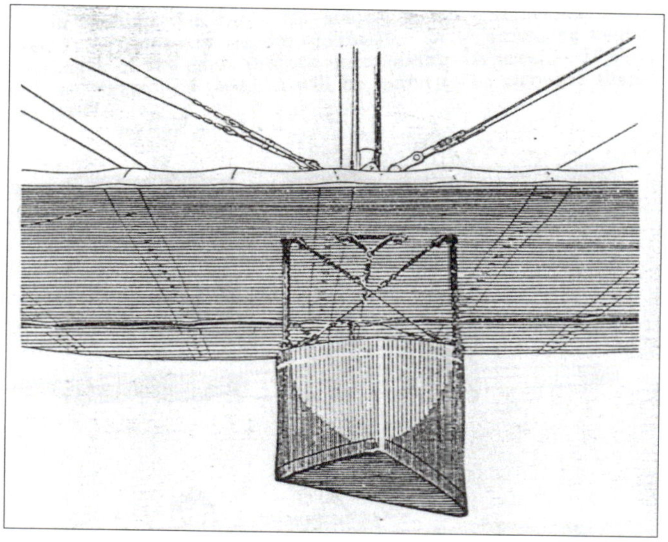

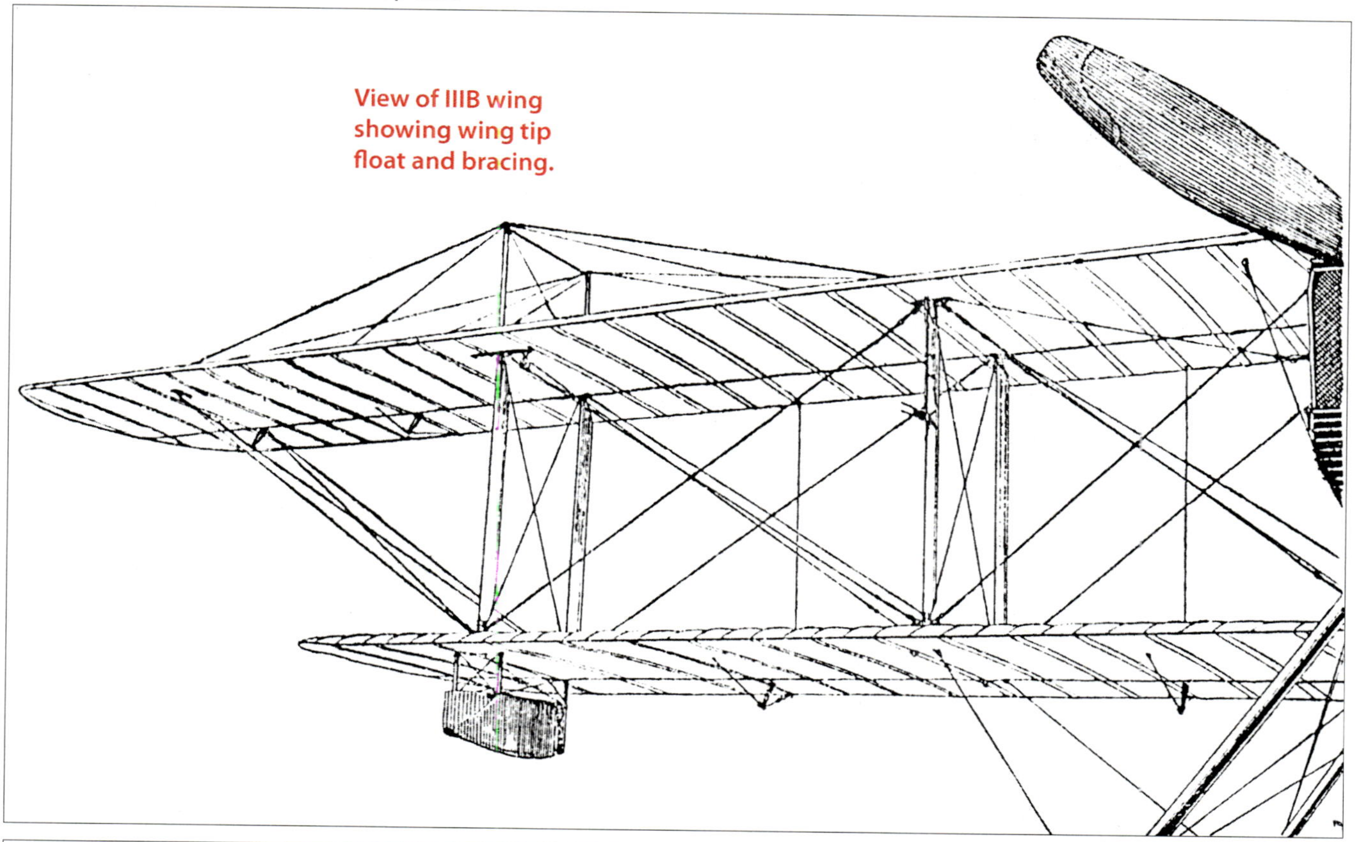

View of IIIB wing showing wing tip float and bracing.

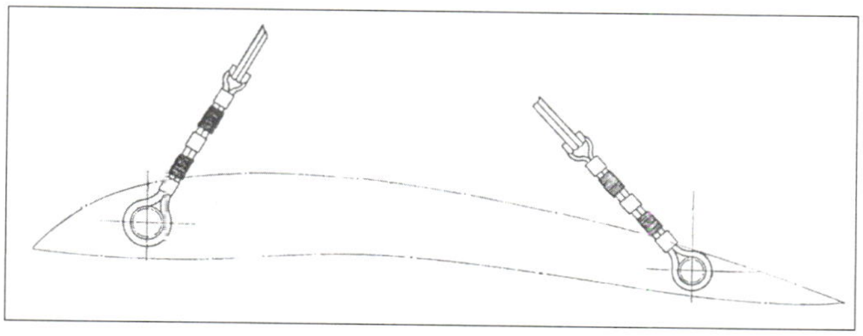

Above: Detail of detachable sling on Fairey IIIC, attached to the centre-section.

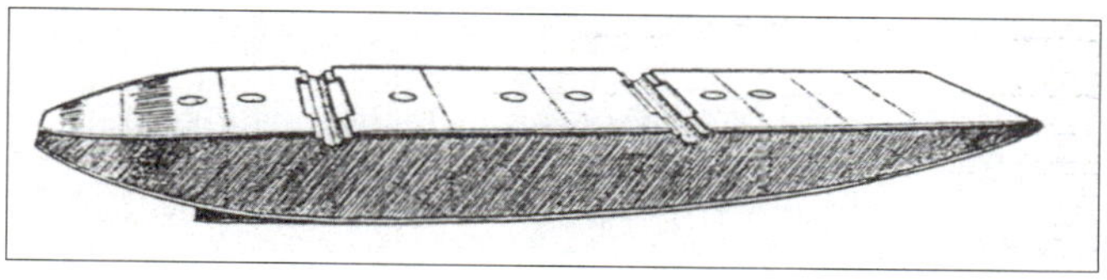

Above: Details of the main floats. The float was 17 ft 10 in length; 2 ft 103/8 in in breadth, and 1 ft 8 ½ in in depth. A single step about 4 inches deep was provided about one-fifth of the length of the float. The inspection covers had air holes to adjust the pressure as the aircraft changed altitude.

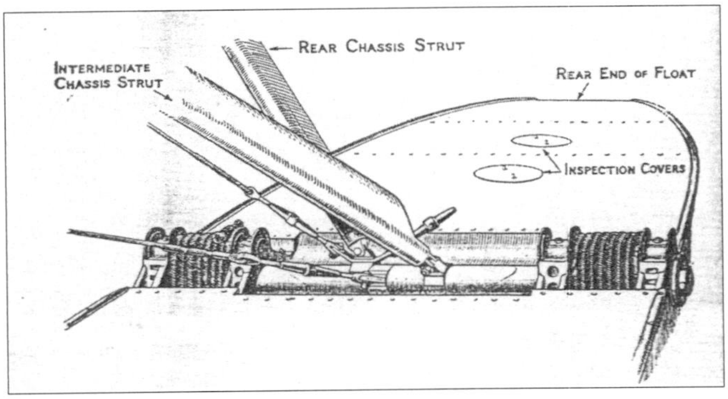

Above: Details of the main floats.

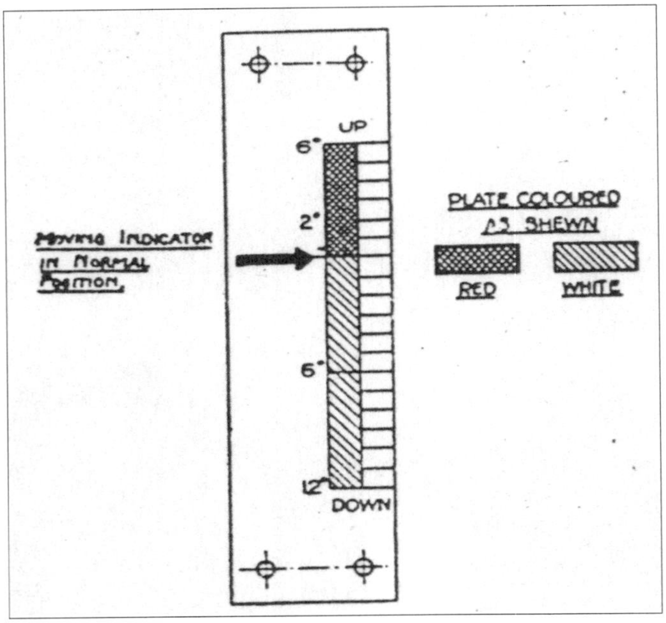

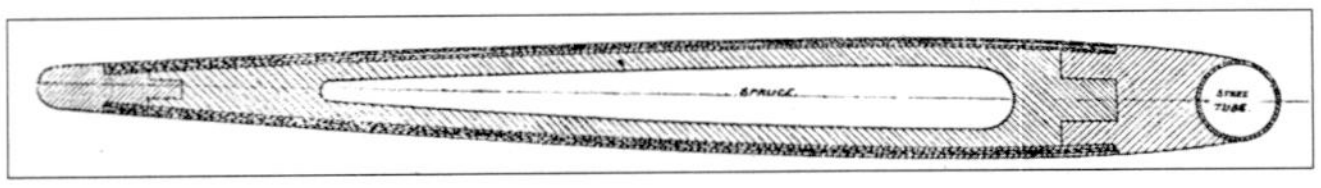

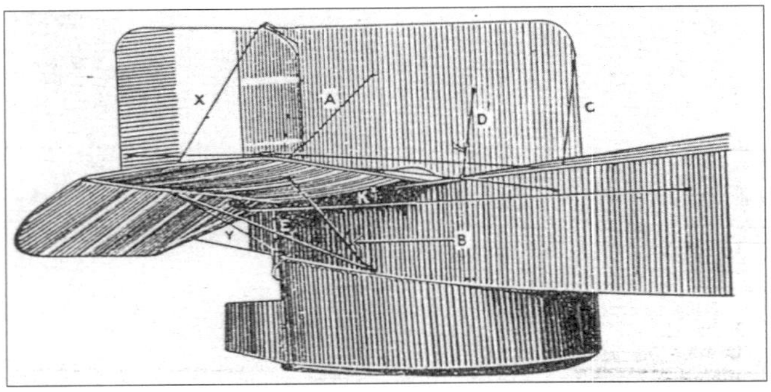

Above: The tailplane showing the arrangement of the bracing. There were two cables on the upper surface and three bracing cables on the lower surface. "K" on the drawing indicates the position of the adjusting screws (one on each side) that may be used to raise or lower the front spar of the horizontal tailplane. Once the incidence was correct (1½° for the IIIB and 2°½ for the IIIC) the adjustment screws were locked.

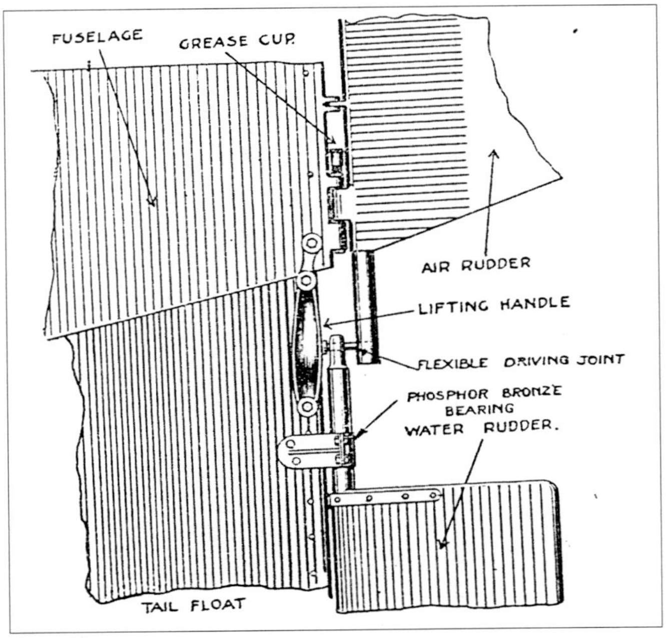

Above & Below: The tail float was built up close to the fuselage underside. The water rudder was attached and was connected to the air rudder as shown.

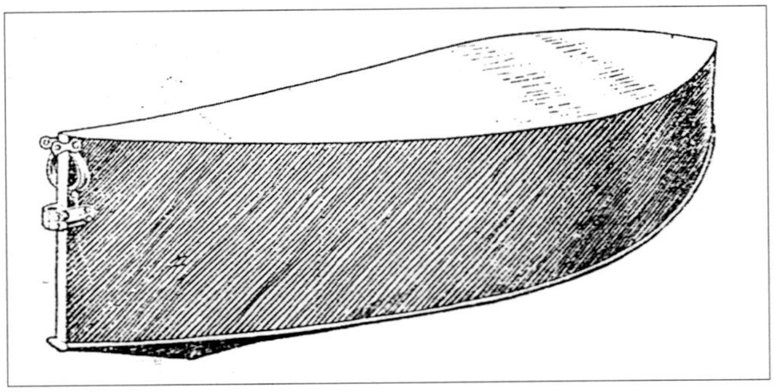

Left: Cross section of the water rudder. The water rudder had a spruce framework covered in three-ply, its front edge was a one-inch steel tube. The rudder was water-tight when completed.

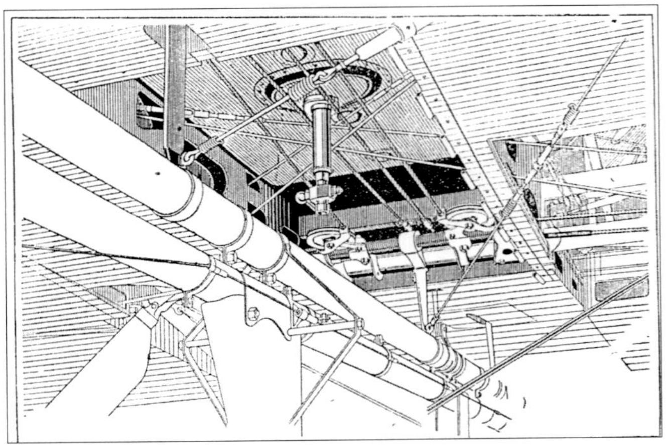

Above: Arrangement of the Camber gear under the fuselage.

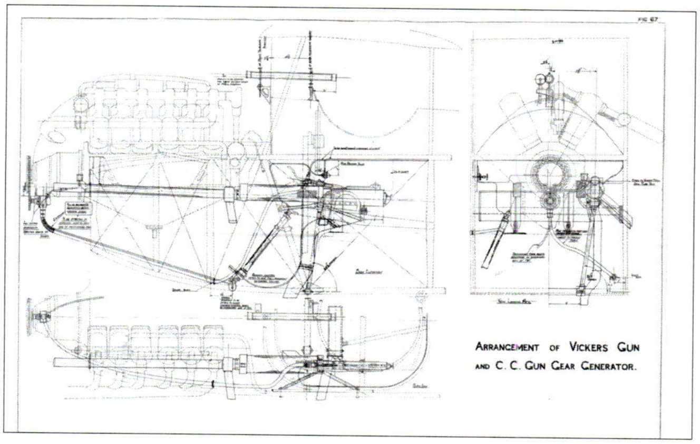

Above: Vickers gun arrangement in Fairey IIIC.

Above: Both the IIIB and IIIC could be fitted with bomb frames below the fuselage. A 112-lb or a 230-lb bomb could be carried, the former in the rear bomb frame by means of special collars and supports.

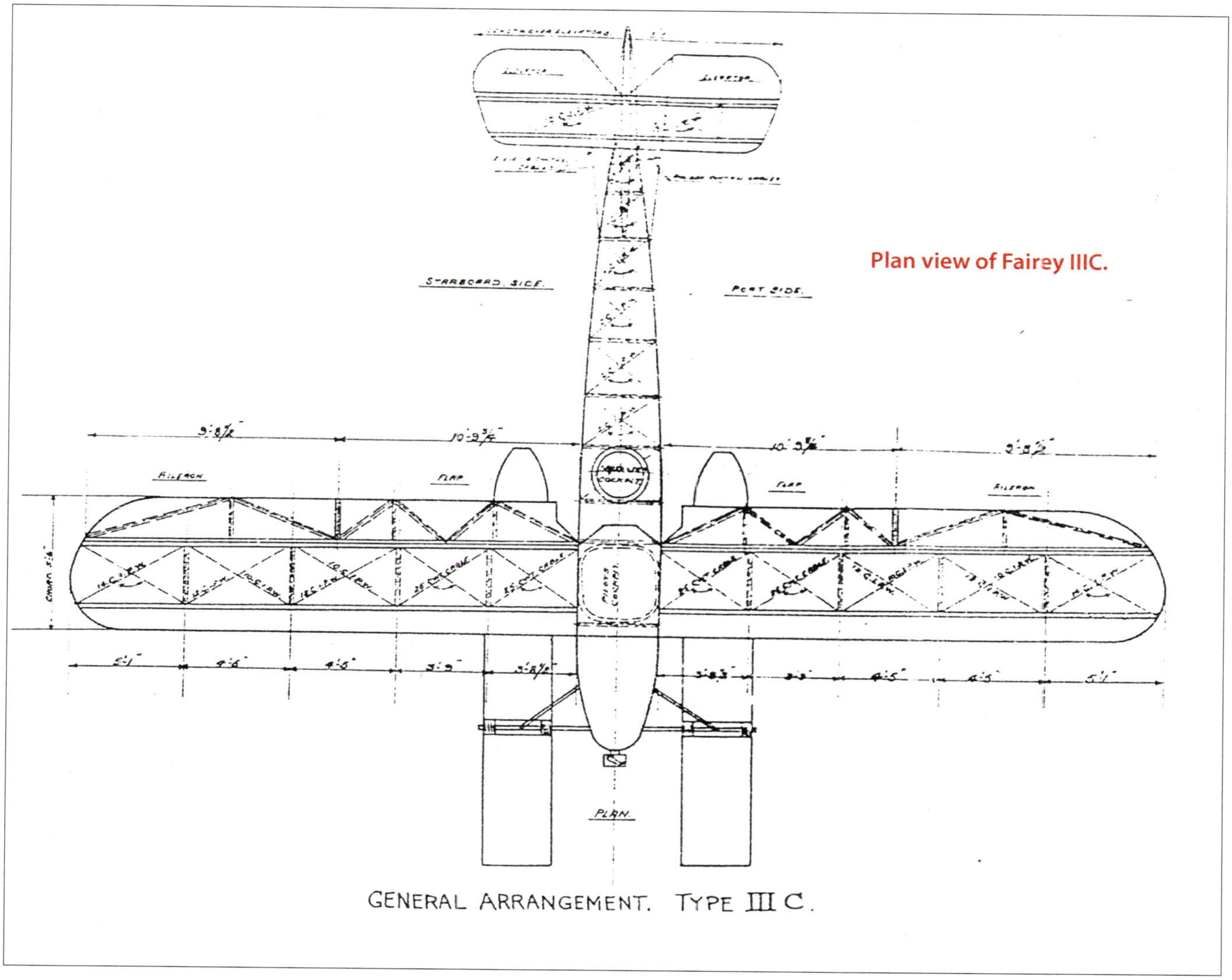

Plan view of Fairey IIIC.

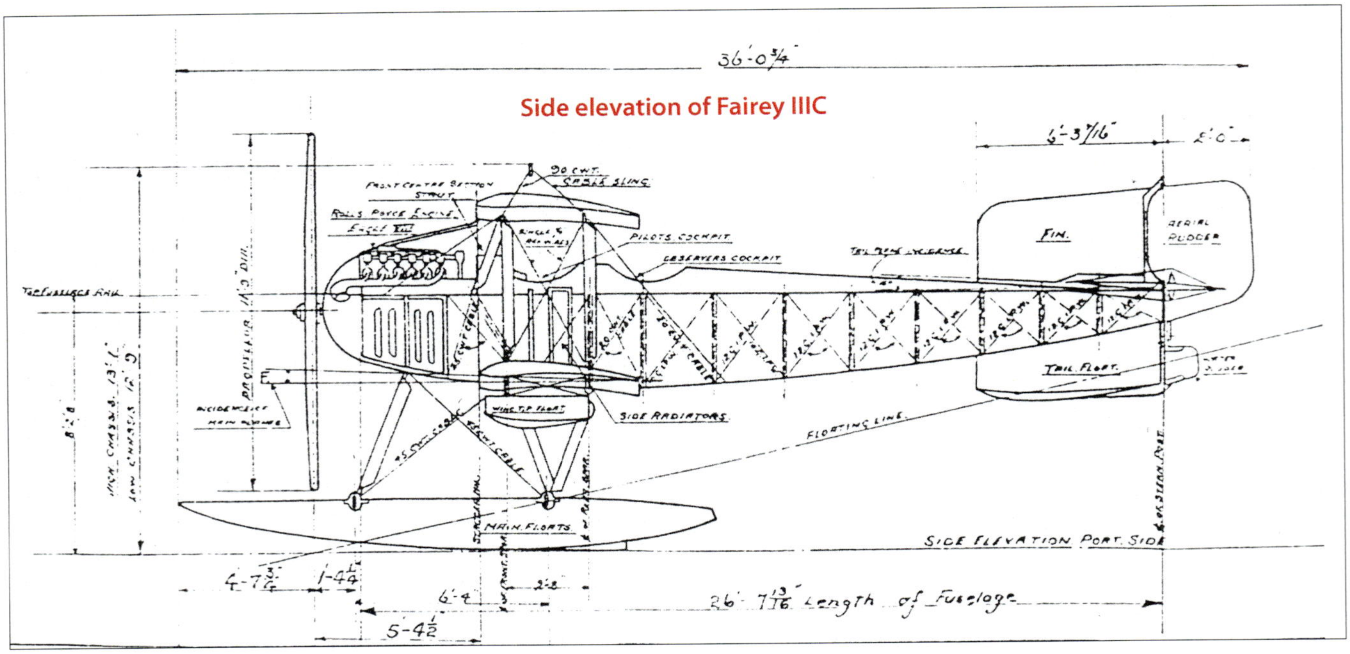

Side elevation of Fairey IIIC

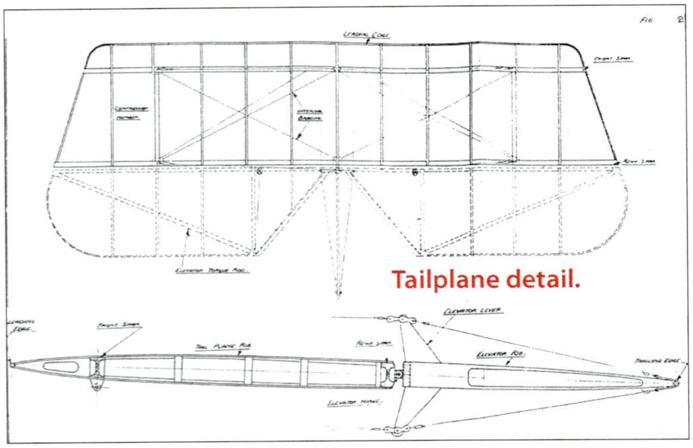

Tailplane detail.

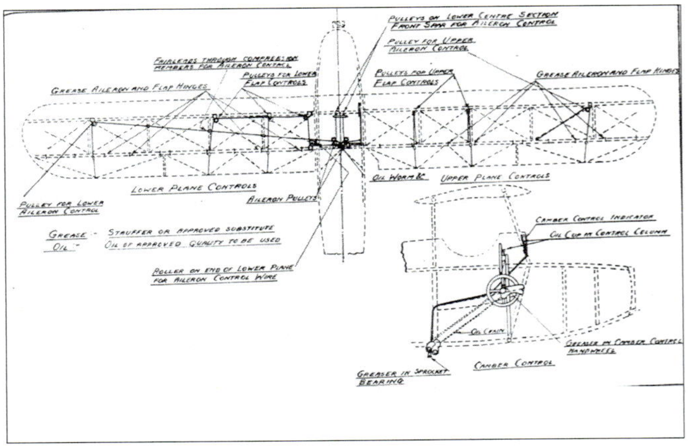

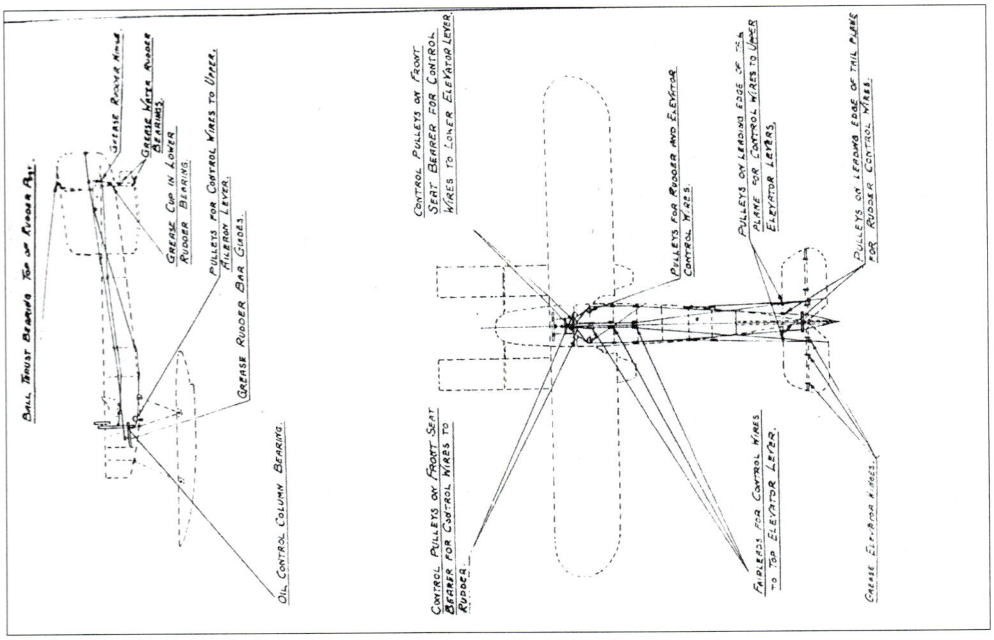

Above & Left: Lubrication chart for tail controls, aileron and Fairey Patent Camber gear controls.

Sketch showing wing folding of IIIC.

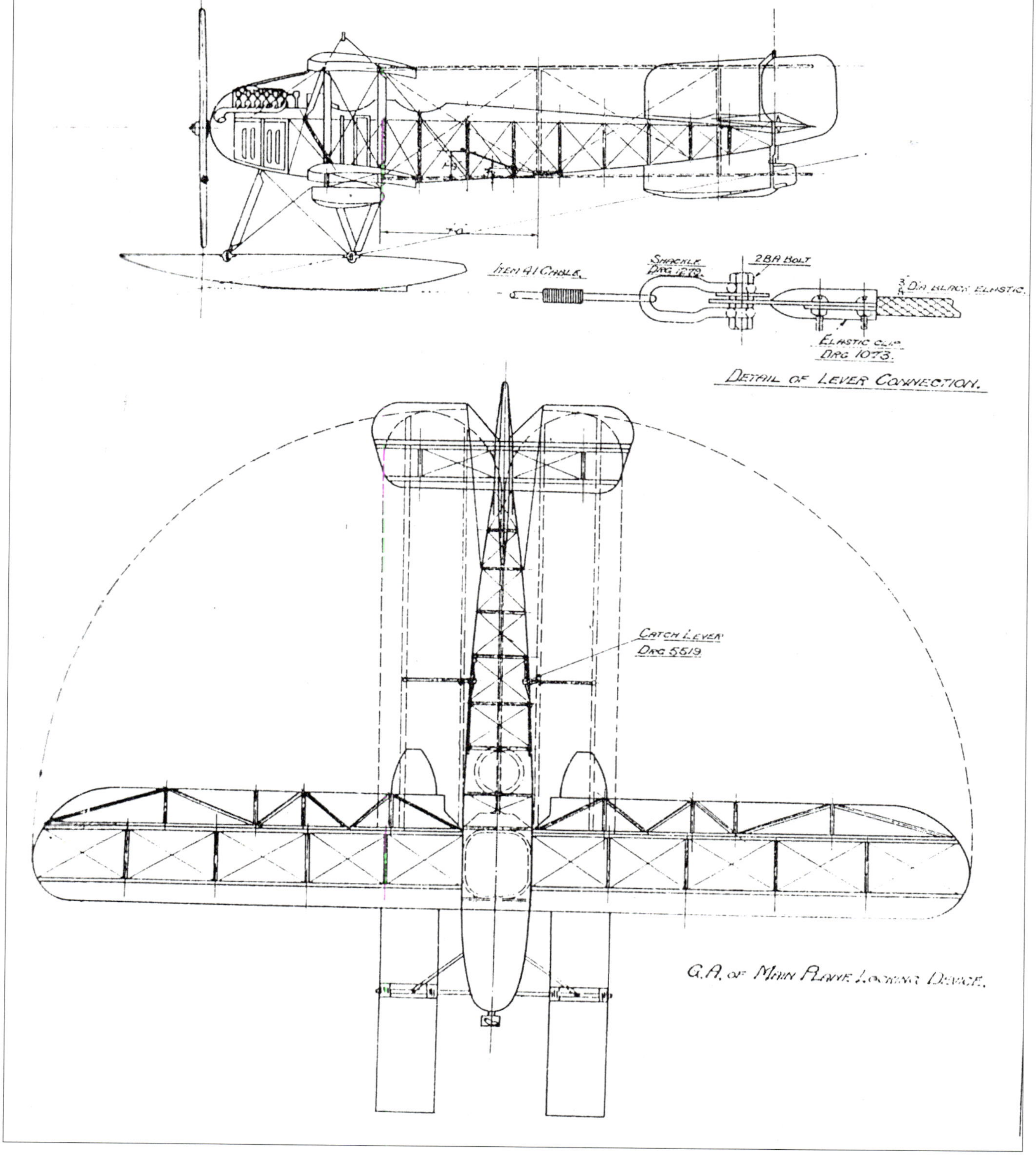

holes to adjust for pressure when flying.

The tailplane was carried over the top of the fuselage and had its front spar connected to the side strut by an adjustable eyebolt. The rear spar rested on the top longerons and was attached to the stern post fitting by an adjustable mild steel hinge. The tailplane could be rotated about its rear spar and be locked in any position by an eyebolt on the rear spar. The incidence of the tailplane depended on the disposition of the load carried. It was not adjustable in flight.

The fin was a tubular steel tube stricture roughly rectangular in shape. The lower surface of the fin passed over the tailplane allowing sufficient clearance to allow full travel of the latter. The balanced rudder was made up of steel tubes braced together by 16 SWG wires.

The elevators were attached to the rear spar of the horizontal tailplane. Both elevators were attached to the same hinge member and were cut away to allow a travel for the rudder of 25° on each side. The trailing edge of the elevators was a steel tube that was carried around to meet the front wooden hinge member. The tail float, when fitted, was built up close to the underside of the fuselage and was constructed of an ash frame with three-ply covering, the compartments being separated by waterproof canvas. The bottom of the float was of ship-shape and had a projecting keel at the rear. A watertight water rudder comprised of a spruce framework with three-ply covering was attached to the rear of the float and had the same movement as the air rudder.

The early IIIA type seaplane was designed for deck launching and did not feature a tail float, and had in place a tail skid with shock absorber. The skid shoe had a ball type fitting to suit the ship's fitting. The IIIA could be converted to a float type "at any time," but the IIIB and IIIC seaplanes were not provided with fittings to enable the fitting of deck launching gear.

The IIIB seaplane was fitted with the Sunbeam Maori II engine that had four magnetos, outside carburettors and inside exhaust pipes. Two oil pressure gauges were provided in the pilot's cockpit.

The IIIC seaplane had the Rolls-Royce Eagle VIII V-12 that developed about 350-hp at 1,800 revolutions. A pressure gauge for the high-pressure oil system was mounted at the port bottom of the pilot's dashboard.

Both types had provision for a No. II Scarff ring carrying a Lewis gun. Stops were provided to prevent the gunner hitting any part of his aircraft. The aluminium covering around the top of the cockpits was dished out to allow easy removal of the gun ring. The gun was fitted with a Norman wind vane sight and provision was made for five ammunition drums to be carried at the back of the cockpit.

The IIIC also had a fixed synchronised Vickers gun on the port side of the fuselage mounted at an upward slope of 3°. A 3 ft 9 in long steel extension tube from the muzzle of the gun provided the pathway for the bullet. A blast tube was fitted at the rear of the gun tube where it was connected to the gun. The ammunition was carried in a belt of 800 rounds in a box across the front of the pilot's seat. The gun was operated through a Constantinesco timing gear. Spent cartridges passed down a chute and out through the cockpit floor. An Aldis sight was mounted over the top of the pilot's instrument board, the supports for which also carried a ring and bead sight.

Both the type IIIB and IIIC were fitted with similar bomb gear to carry two 230-lb bombs. The bomb frames were carried on the bottom of the fuselage, one in front of the other. *Great care must be taken in fitting the bomb gears as a special clearance is necessary in order that the bombs may not foul the rear cross boom on the floats when they are released.* A Mark II.A low-altitude drift bomb-sight was mounted by means of brackets on the starboard side of the gunner's cockpit.[84]

The wireless gear comprised a windmill-driven generator mounted on a swivelling bracket in the starboard side of the fuselage. When not in use the whole unit could be swung inside the fuselage and covered by an aluminium door. Both aircraft had the following instruments, but the arrangement of panels was different in each type. These instruments comprised:-

Type 57 transmitter.
Model T.F. receiver.
Variable resistance for generator field.
High tension voltmeter.
Tumbler switch for Aldis lamp.

In the IIIB the instrument panels were arranged vertically across the fuselage midway between the two cockpits. In the IIIC the instrument panels were placed on the port side of the gunner's cockpit. A locker for spare wireless parts was located at the rear of the gunner's cockpit on the port side near the ammunition rack.

The aerial comprised 120 feet of flexible copper wire 1/16 in in diameter. A 2-lb weight was attached to its lower end. This was lowered after the machine had reached an altitude of at least 200-300 feet.

The Type III aircraft were "constructed to be aero-dynamically stable" so that they would maintain a given flight path with all controls released at cruising speed, i.e. between 75 and 85 knots. The engine was offset 1° and it was important "in changing aircraft engines to note that an engine of the same direction and rotation must be installed, failing which fresh bearers must be provided to give the corresponding 'off-set' in the correct direction."

When manoeuvring on the water the wing flaps were to be fully wound up to prevent the machine rising prematurely. The aircraft was very easy to manoeuvre at low speeds with the water rudder immersed. Lateral control was extremely quick and sensitive in a properly erected machine.

Pilot's Opinion of the IIIC

In his magnificent book on the history of the Great Yarmouth Seaplane Station during the war (*The Story of a North Sea Air Station*), C.F. Snowden Gamble wrote that little progress had been made in 1917 with the design of float seaplanes. The *Fairey Aviation Company produced the IIIc float seaplane*

The Normal Load Was Worked Out as Follows

Item Description	lbs
Machine light including Scarf ring, W/T fittings, bomb sight, & bomb frames	3,437
Lewis gun & 3 double trays.	45
Very's Pistol & 12 cartridges.	9
W/T transmitter (Type 52.N) & T.F. receiver complete.	70
Crew	360
Water 10½ gals	105
Petrol for 3 hours (70 gals).	490
Oil for 4 hours (4 gals).	40
Aldis lamp & battery.	20
Emergency rations, Medical bag, Tool kit & coil rope in haversack.	15
Crew's revolvers & Ammunition	12
2 x 112-lb bombs	224
2 x Cooper bombs	48
TOTAL	4,875

(a machine of this type was not issued to Great Yarmouth air station until November 1918), which was a distinct improvement upon either of its predecessors. The three-float suspension system with wing-tip floats was employed for the undercarriage. The mainplanes were arranged to have equal span for the upper and lower wings, unlike all previous Fairey types. *As with the Campania, the radiators for the cooling system were arranged on the sides of the fuselage, but placed much further aft, being in line with the rear of the pilot's seat.*[85]

B. G. *Guy* Blampied flew the IIIC in Russia. He found them to be *extremely nice seaplanes to fly. I thought a big improvement on the Fairey Campania which I flew when I was in "Campania." I only flew Shorts when training at Calshot & Lee on Solent. I thin the Short 150 hp & Sunbeam 230 hp. I also flew Shorts at the Isle of Grain … but do not remember much about them.*[86]

The Report on the use of the IIIC in Russia recorded that the Fairey IIIC *is by far the best Seaplane of its type yet produced in Great Britain.* The IIIC seaplanes were flown up river on various dates from 16 June onwards. From that time onward, they never had any protection from wind, rain, sun or sand other than that afforded by engine and cockpit covers. The rainy season proper had arrived, and on several occasions, heavy storms with tropical rain of great force were experienced. The procedure was to beach the seaplanes on the sandy beaches, some steep, others with gradual slopes, with their tails to the beach and the heel of their floats aground.

Despite the lack of supplies and the lack of time to make complete overhauls of the machines, the skill of the pilots and the great devotion of the mechanics, who worked day and night to keep the maximum number of seaplanes serviceable, saw the Fairey seaplanes stay in operational service. Some floats were reconstructed up to seven times.

Despite the permanent exposure and climatic conditions, these seaplanes have kept in very good condition. No warping has occurred in fuselage or wings, and the whole machine has kept in truth remarkably well.

Fabric has kept very well; it is taut and there has been no "peeling" of dope.

The Fairey IIIC *requires knowing and pilots* (are) *require to fly it for 3 or 4 hours before flying her at her, or their best.*

She is difficult to land, and must not be loaded above 4900 lbs, but at this load when can be handled as easily as a land machine.

At the maximum load of 5050 lbs. the Seaplane is heavy and unhandy and feels overloaded.

The load of two 230-lb bombs was too much. All the machines on the Russian intervention were fitted to carry two 112-lb bombs and from two to three Cooper bombs that were carried by the observer. The Vickers guns were removed as they were not needed as aerial fighting did not take place and the observer's guns were better for strafing.

The normal load was worked out as shown in the table. With this load the seaplane got off well and had a good performance. At full throttle she would fly at 85 to 90 knots. *The adjustable flaps are invaluable.*

The best position for getting off, climbing and flattening out for the last 50 feet of the glide is 4 degrees down – when flying level, whether "all out" or throttled the FLAPS at 0 degrees down increase the speed. If any up angle is given, however, the machine loses height and it is not considered, that it is of any use having them capable of being put at an "up angle."

The seaplane will achieve flight at 42 knots, climb at 42 knots without stalling, but the best climbing speed is 50 to 53 knots.

Her glide is not good (and in this respect she resembles the Fairy Campania Seaplane). If glided below 60 knots she lose height, not on the line of glide but vertically, and if her speed is not increased before the last 50 or 60 feet, she continues her "semi-pancaking" motion, quite unaffected by her elevators; any movement of the latter merely altering the angle at which her floats hit the water, and not her angle on the gliding path. Her best gliding speed is 60 knots, when she exhibits none of the above tendencies, but glides normally although the angle is not a good one.

This seaplane turns far better at a steep bank than if turned at about 10 degrees of bank, it being almost impossible to stop "slipping" is turning with small bank. When banking steeply full "up elevators" are required and top rudder, and the bank requires holding off a little.

Having brought the Seaplane down at a gliding speed of 60 knots (with flaps at 4 degrees down as previously stated) she will land normally, but at a fast speed.

Her "loading" being high, she loses speed very quickly and will "pancake" badly if held off the water too long. The difficulty of landing the *Fairey 3/C* well and consistently due to her heavy load per square foot of wing surface was commented on elsewhere. During June and July there were frequently no surface wind and the glassy calm of lake rendered good alightings extremely difficult to make. The lack of spare floats added to the delay of returning the machine to service when there was minor damage.

The Seaplane has extremely good lateral and fore and aft control (except as mentioned in the paragraph re "Gliding."), but the directional control, although ample, is slower in proportion to the lateral control. This is probably due to the large fin, but it has not been possible, so far to carry out any trials with a smaller fin.

In "bumpy" weather her behaviour is good.

The floats have stood continual immersion and the sun extremely well, but an improvement that has been fitted on them, and which prevents the upper three-ply from warping, is the addition of two slats of wood (3½" wide by ½" thick), screwed on the top of each float, from heel to nose. This allows men to stand on the floats without risk of penetration, and also prevents the top of the floats from wrinkling and strengthens them.

The floats are of sufficient strength and of a very good design, enabling the Seaplane to get off the water very quickly; unless realised and prevented, the machine is liable to "porpoise" when getting off and landing, and has to be "Piloted" off the water (in contradistinction to the "Short" Seaplane 184 Type which will practically take itself off). The single step is in my opinion the obvious solution to the difficulty of the twin float design, and the contrast between the Fairey 3.c. and the Short 184 type floats, hydroplaning and "unsticking" is most marked.

There has been no opportunity of judging the seaworthiness of this Seaplane, for the Dwina River is always calm, but, though she would probably get off on a rough sea, her landing speed is too great to make a landing on a rough sea, anything but a gamble. If, however, landed satisfactorily, it is considered that she would ride well, but one would be well advised to drain all petrol and lighten, as much, and as quickly, as possible.

The Seaplane taxies remarkably well and has great controlability with the water-rudder; no difficulty is experienced in turning out of a wind.

The pilot's view is excellent, the Observer's is good except for Bomb dropping from a low height, when the trailing edge of the lower plane comes in the way.

Structural Defects and Suggested Improvements:

The rear upper undercarriage strut attachments, which are on the fuselage, are not of sufficient strength, and they should be made of high-tensile steel instead of cast as at present. Two Seaplanes have fractured both Port and Starboard of these lugs through landings, (which, though admittedly bad and resulting in crashed floats and struts, should not have fractured these lugs), for sweated as they are on to the athwartship steel tube which forms one of the fundamental parts of the fuselage, their replacement can only be effected after stripping the fuselage, removing the engine, and practically dismantling the whole machine.

The Vickers Gun box was in the way of the pilot's legs and was not conductive to his comfort. It was removed from the IIIC machines in Russia as noted above. It was suggested that the idea of carrying 230-lb bombs *should be discarded* for bombing operations and all machines fitted to carry two 112-lb bombs and a special rack for four Cooper bombs be fitted on the top of the rear after petrol tank in the observer's cockpit.

No mention has been made of the engine, which proves itself, once again, to be the most efficient stationary engine yet produced.

This report appears to have been written by Lt Col Lancelot Talison, RAF. While full of praise for the IIIC it did not give any information on the actual use of the IIIC seaplanes in action.[87]

End Notes

1. Copy in Smithsonian NASM File No.AF-050000-02 Fairey, General.
2. Report No. N.M. 107. 22.02.1918. TNA AIR2/969.
3. Report No. N.F.110. From OC Test Depot, Isle of Grain to Assistant Controller, Experimental & Research Dept, 28.02.1918. TNA AIR2/969.
4. Report N.M.145A. 04 May 1918. Maj N Wright, OIC, to Assistant Controller, Experiments. Also signed by Lt Col H Busteed, CO. TNA AIR2/969.
5. Report No. N.F. 108. *Report on Structural Defects on Fairey seaplane N.10. built by the Fairey Aviation Co.* Test Depot, Isle of grain, 28.02.1918. TNA AIR2/969.
6. Minute A Hayward for Head of D.3.c to D.3. 05.03.1920. TNA AIR2/971.
7. RAE Report No.M.T.5041. Strength test of Main Planes for Fairey 3C Seaplane. (30 Dec 1921) and Test of 24 November 1920. TNA Avia 6/5897.
8. Discussing the Fairey 3.C carry-over production order in February 1920, (this would be for the IIID), it was revealed that with *regard to the drawings for the Fairey 3.C. seaplane, it is desired to point out that this type of machine has never been standardised for production and therefore a standard set of drawings is not in existence. It is proposed, however to construct the new production machines in all respects similar to the Fairey 3.a. seaplanes delivered during the last twelve months for operations in North Russia* but with modifications. TNA AIR2/971.
9. Controller, Technical Dept to Fairey Aviation Co, *Re. N.111*

Type IIIC Specifications

Source	J.M. Bruce Data	Janes 1919	IIIB/IIIC Manual
Span upper	46 ft 1¼ in	46 ft 2 in	46 ft 1¼ in
Length	36 ft		36 ft 0¾ in
Length wings folded			36 ft 0¾ in
Height	12 ft 1¾ in		
Height - High chassis			13 ft 1 in
Height – Low chassis			12 ft 9 in
Chord	5 ft 6 in		5 ft 6 in
Gap	5 ft 3 in		5 ft 7 in
Span of tail	13 ft		13 ft 0 in
Dihedral	1° 04'		1° 40
Floats			
Length			17 ft 10 in
Max width			2 ft 10 in
Max depth			1 ft 8½ in
Wing Area, Sq. ft.	476		476
Aileron Area, Sq. ft.	16.5 each		16.5 each
Tailplane Area, Sq. ft.	34.2		34.2
Elevator Area, Sq. ft.	34.2[1]		11 each
Fin Area, Sq. ft.	12.4		12.4
Rudder Area, Sq. ft.	12.4[1]		15.3
Total Wt, lb.		5,050	4,800
Light Wt. lb.			3,549
Military load, lb.			258
Crew Wt., lb.			360
Fuel Wt., lb.			812
Oil Wt., lb.			71
Disposable load Apart from fuel		1,030	
Fuel Capacity, gall	120		112
Oil Capacity, gall			7½
Endurance, hrs		6	
Speed at sea level		97 kts	
Speed at 2,000 ft			96 kts
Speed at 10,000 ft		90 kts	
Min safe speed			43 kts
Min safe gliding speed			48 kts
Landing speed		44 kts	
Climb to 5,000 ft		6.40 min	
Climb to 10,000 ft		17.30 min	

Notes: 1. These amounts are incorrect.

B Experimental Float Seaplane. 09.04.1918. TNA AIR2/969. Despite this, documents continued to refer to the "Fairey Type 3 Seaplane (N.10)," such as one of July 1918 where in the type's performance is considered when powered by an Eagle VIII, a Liberty and a Sunbeam Maori engine. Technical Report No.344. 19.07.1918. TNA AIR2/969.

10. Minutes Royal Navy Air Department Progress Committee, 20th Meeting 25 September 1917. TNA AIR1/270/15/266/113.
11. Fairey Aviation Co Weekly Report. 21.11.1917. TNA AIR2/969.
12. Weekly Report of Fairey CO for W/E 21.11.1917. Copy in TNA AIR2/969.
13. Memo from T.6 to T.I.N. "Fairey 3A. (No,10) Seaplane. N.2.a. 30.11.1917. TNA AIR2/969.
14. 61st Report of the Controller, Technical Dept, for the fortnight ending 21.08.18. Copy in RAFM J.M. Bruce Collection Box 25. A Smith in *The Life of Sir Richard Fairey – The man who built the Swordfish* (I.B. Tarus, UK, 2018) calls the Type IIIA land-based bombers and not ship planes.
15. Notes from TNA AIR 1/2411/303/4/28, Progress & Allocation Committee Minutes of Meetings, June-Dec 1918. Copy in RAF Museum JM Bruce Collection Box 6.
16. JSB for Lt Col A-C.T.D.(Design) to X-4. *Fairey Ship Aeroplane*. 13.06.1918. The hydrovane gear as designed by Fairey was approved around June 1918. TNA AIR2/969.
17. E.R. Fairey, Sole Managing Director, Fairey Aviation Co Ltd to The Controller, Technical Department Aircraft Production, (For the Attention of Maj Miley). 03.09.1918. TNA AIR2/969.
18. DAD Admiralty to Air Ministry, 22.07.1918. TNA AIR1/30/15/1/154.
19. Fairey Seaplane Makers Type N10 200 HP strength. Report of 29.11.1917. AIR2/969.
20. Fairey Aviation Co Weekly Report. 21.11.1917. TNA AIR2/969.
21. "Forty-Fifth Meeting of the Technical Committee" 30.05.18. Copy in US NARA RG72 File O-ZS-7.
22. All quotations in this section are from Dept of Aircraft Production Minute, signed A.J. Milley(?), for Lieut Col, RAF, Assistant Controller, Design. 19.07.1918. TNA AIR2/972.
23. Minutes of Conference held in Air Ministry 31.08.1918, to consider the whole question of the relative suitability of the boat and float seaplane for anti-submarine and other duties. Copy in RAFM JM Bruce Collection Box 3. In early 1919 the Seaplane Experimental Station at Felixstowe was asked to complete the comparative testing between the D.H.9 with floats and the Fairey IIIC. In reply it was noted that the work on the D.H.9 seaplane had been stopped due to work with the proposed trans-Atlantic flight and the shortage of staff. It was not considered that *the D.H.9 machine on floats will show any advantage over the Fairey 3C for reconnaissance purposes and the latter should be more seaworthy and more reliable in every respect as a seaplane*. TNA AIR2/970.
24. Dept Aircraft Production Minute addressed to Maj Penny by Maj Gnosspelius. 29.07.1918. TNA AIR2/972.
25. Dept of Aircraft Production Minute, submitted to Maj Penny by O.T. Gnosspelius, Maj RAF. 29.07.1918. TNA AIR2/972.
26. Hand-written note by Major Gnosspelius, 19.08.18. TNA AIR2/972.
27. Memo: Major Gnosspelius to Lt Col Beatty, 22.08.18. TNA AIR2/972.
28. Notes by J.M. Bruce from TNA file AIR 1/500/15/331/4 Dept of Aircraft Production – Reports 79, 80 & 84. Sept-Oct 1918. RAF Museum J.M. Bruce Collection Box 7.
29. Technical Dept to Ministry of Munitions in Paris. 26.09.1918. TNA AIR2/972.
30. R Penny for Controller, Technical Department to Ministry of Munitions, Paris. 03.10.1918. TNA AIR2/972.
31. Dept of Aircraft Production Ministry of Munitions, Paris to Controller, Technical Dept, Air Board. 05.10.1918. TNA AIR2/972.
32. Capt Eric Neal for Director DAP, Paris, to Controller, Technical Dept, Air Board. 11.10.1918. TNA AIR2/972.
33. Handwritten note to Minute 29.10.18. TNA AIR2.972.
34. "Cooling System for Fairey Seaplane Type 3.C." Memorandum from Ap.D(E) to Ap.D(S). 23.10.1918. TNA AIR2/972.
35. "Cooling System Fairey 3C Seaplane with Eagle VIII Engine". Lt C Anderton Brown, RNVR, for Ap.D.(E.4) to Ap.D.(s). (Attn: Capt Goodwin). 03.02.1919. TNA AIR2/972.
36. Dept of Aircraft Production Minute: *Tanks for Fairey III C seaplane*. To Maj Penny by O.T. Gnosspelius, Maj RAF.31.10.1918. The Test Depot on the Isle of Grain reported that the baffle plates in the tanks were not notched or drilled at the top thereby causing an air lock. The capacity of the tank was 51 galls but only 41 could be put in. After the air lock was overcome the total was 28 galls. Grain Test Report No. NF236. 01.11.1918. TNA AIR2/972.
37. Another memorandum indicated that Maj Wright has been carrying out flying trials of the prototype IIIC. Assistant Controller Design to Ad.D. 05.10.1918. TNA AIR2/972.
38. Minute: *Experimental light of 2 seater Fighting Seaplane*. Maj Wright. 02.10.1918. A Table of Machines to meet the RAF Type XXXI Float Seaplane Load 2.B. was prepared on 25 July 1918, comparing the RAF Specification with the proposed performance of the Fairey Type 3 with the Maori, Eagle VIII and a Liberty engine. TNA AIR2/972.
39. Handling Report Grain NM 238. TNA AIR1/35/15/1/218.
40. *Grain Test Report NF244 on Petrol Tank of Fairey No. N.2246*, from Capt CE Moore to Assistant Controller, Experiments. 21.01.1919.
41. Minute. Maj A.J. Miley, Dept Aircraft Production to Lt Col Stedman, 08.02.1919. TNA AIR2/970.
42. Minute: "Fairey IIIC conversion from IIIB". Head of DC to An.D (8). 07.03.19. TNA AIR2/970.
43. D.D. of N.(F) to D.A.S. (Mr Wilson). A note "By Hand"

was added to the file copy. 06.05.1919. TNA AIR2/970.

44. F.M. Charles, General Manager to Capt A.J. Goodwin, A.P.D.(S). 07.03.1919. TNA AIR2/972.
45. Gen Courtney, Director Aircraft Equipment, to Controller Technical Dept, Ap.D. 16.04.19. TNA AIR2/972.
46. This is an error in transcribing the original report. The IIIC would have been N9250 that was with *Pegasus* at this time. N5250 was an Avro 504B trainer.
47. L.T., OC, RAF Units, HMS *Pegasus* to Air Ministry, Memo: *Weakness of Radiators on Fairey 3C Seaplane.* 24.05.1919. UK AIR2/972.
48. A. Courtney, DDE to Director of Research, 11.06.1919. UK AIR2/972.
49. The number of aircraft delivered by *Pegasus* differs in sources. All agree that the number of Fairey floatplanes was approximately twice that of the Short 184 floatplanes. Despite what is sometimes written, only the Fairey IIIC version of the Type III served in Russia. When the IIIB were re-engined they emerged as IIIC seaplanes.
50. Lt Col R.J. Bone's Report to General Officer Commanding North Russia Expeditionary Force, 1 October 1919. TNA ADM137/1713.
51. "Resume of Operations and Events for Period 1-20 April 1919". Lt Col Robin Grey, Acting C.RAF, Allied Forces, RAF General HQ, Archangel, to Air Ministry. 01.05.1919. TNA AIR1/435/15/274/3.
52. Livock GE, *To the Ends of the Air,* HMSO, UK, 1973. PP67-73. Livock served from October 1914 in the RNAS flying Short floatplanes before becoming a "Big Boat" pilot. He remained in the RAF and worked on the development of long-range reconnaissance flying boats. He led the Far East Flight of Supermarine Southamptons to Australia in 1928. From 1940 to 1941 he joined the staff of the newly formed Coastal Command. He retired as Group Captain and was headmaster at a boy's school from 1946 to 1954.
53. Lt Col R.J. Bone's Report to General Officer Commanding North Russia Expeditionary Force, 1 October 1919. TNA ADM137/1713.
54. "Log of Flights." Hand-written report in Army Book 129. TNA AIR1/450/15/312/9.
55. "Log of Flights." *Ibid.*
56. Quoted in Tomaselli, P. "Faireys over Lake Onega. Air Operations with RAF Syren Murmansk 1919," *Cross & Cockade International*, UK, Vol. 28 No. 3, 1997. P.133.
57. Letter. *Guy* Blampied to P Thomaselli, April 30, No year (probably 1990). P Thomaselli Collection.
58. Tomaselli, P. "Faireys over Lake Onega. Air Operations with RAF Syren Murmansk 1919," *Cross & Cockade International*, UK, Vol. 28 No. 3, 1997. P.135.
59. Letter. *Guy* Blampied to P Thomaselli, No date. P Thomaselli Collection.
60. Message to HAIR, from OC Duck Flt. No date. The seaplane flight was called the Duck Flight. TNA AIR1/472/15/312/167.
61. "Extracts from Report of Operations S.N.O.(R)." TNA AIR1/16/15/1/81.
62. Lt Col R.J. Bone's Report to General Officer Commanding North Russia Expeditionary Force, 1 October 1919. TNA ADM137/1713.
63. Khairulin. M. *Russian Aviation Colours 1909-1922, Vol.4 Against Soviets.* MMP Books, UK. 2018. P.87.
64. Notes from the late H Woodman.
65. Weekly *Disposition of Aircraft* Reports. 05.12.1918. TNA AIR1/670.
66. "The Schneider Cup Race". *The Aeroplane*, 27 August 1919, P.802.
67. It was common practice for former servicemen to be addressed by their rank in British aviation journals and newspapers of the day.
68. Fairey Aviation CO to Technical Controller, Air Ministry. 18.09.1919. TNA AIR2.970.
69. Joseph George Navarro purchased N2876 from the Disposal Board without engine and on delivery it was found that it was not fitted with floats. AIR2/970.
70. The file ends here. It is unknown if the two machines were granted a CofA before they flew to Sweden. TNA AIR2/970.
71. Fairey Aviation Co to Director of Research, Air Ministry, 20.01.20. TNA AIR2/971/286/45 Pt.III.
72. *The Aeroplane* 28 Jan 1920. P/226. Taylor states that two Fairey IIIC floatplanes were shipped to Sweden by Henderson, however the journals of the time mention three.
73. "The Activities of the P.O. Flygkompani." *The Aeroplane,* 3 March 1920, P.498.
74. "Sweden." *The Aeroplane.* 17 March 1920. P.594. In "Master of the Air, Jimmy Youell", P Youell states that Jimmy had a case of snow blindness and this caused his crash, his only crash until his death on 19 April 1961, in a Bell 47J-2 helicopter. Bruce *Jimmy* Youell joined the RNAs and there he met Col G.P.L. Henderson. *Aeroplane,* March 2001, P.38.
75. Mulder, R. "The Most Northerly Aerial Service of the World: Porjus-Suorva (1920-1923)". The three Avro 504L floatplanes were S-IAA, S-IAB and A-IAG.
76. Letter: Fairey Aviation Co to The Controller, Technical Dept, Aircraft Production, RD: *Atalanta, Campania & Hamble Baby.* 14.01.19. TNA MUN8/3.
77. Isle of Grain Report for f/night end 21 December 1918. Seventh Report of the Controller, Technical Dept. for week ending 01.01.19. RAFM JM Bruce Collection Box 25.
78. N2246 was the only Fairey IIIC still in service after 01.01.1920 according to Thompson, D & Sturtivant, R. *Royal Air Force Aircraft J1 -J999 and WWI Survivors,* Air Britain, 1987. This conflicts with other soyrces.
79. WA Robertson to Fairey Aviation Co, *Fairey 3 C. Seaplane for flight from London to the Cape.* 08.01.1919. This letter was apparently lost and it was resent on the 15th.TNA AIR1/35/15/1/218.
80. C.R. Fairey to Air Ministry. 16.01.1919. TNA AIR1/35/15/1/218.
81. TNA AIR2/972.

82. Mason, F.K. *The British Bomber since 1914,* Putnam, UK, 1994.
83. *The Fairey Seaplane Types III.B and III.C.* Air Ministry, 1921. Copy at TNA AIR10/410.
84. On February 1920, the armament of the IIIC was discussed a Meeting in the Air Ministry. It was stated that no schedule of armament for the IIIC had been issued. It was decided that *each machine should be fitted with an ordinary tubular bomb frame to which it must be possible to attach either a single 230 pounder tubular bomb rack or a double 112 pounder tubular bomb rack.* The Course Setting Bomb Sight Mk.I was to be fitted, the actual release gear being fitted by the unit according to the types of bombs it is proposed to use. The observer was to have a Lewis Gun with 100/100 Norman vane sight, on a Mk.II Scarff ring. *The question of fitting a Vickers gun to the machine was to be taken up and a decision will be given as soon as possible.* In order to avoid any confusion on the matter of spares with the original F.3.C. type the machine was to be known as the F.3.D. TNA AIR2/971.
85. Snowdon Gamble, CF. *The Story of a North Sea Air Station,* Oxford University Press, 1928. P.290.
86. Letter B.G. Blampied to P Tomaselli, April 30. Year not stated. P Tomaselli Archives.
87. "Report on the Fairey 3.C. Seaplane, with Eagle 8 Rolls Royce Engine, on Active Service on Dwina River with B.N.R.E.F. Summer 1919." OC RAF Dwina River Force to Air Ministry, 19.09.1919. TNA AIR2/970.

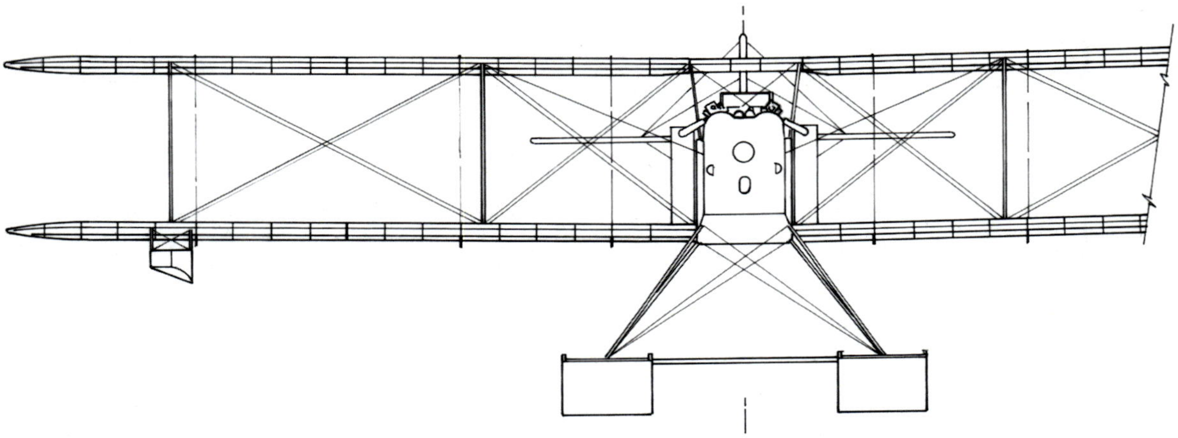

Fairey IIIC

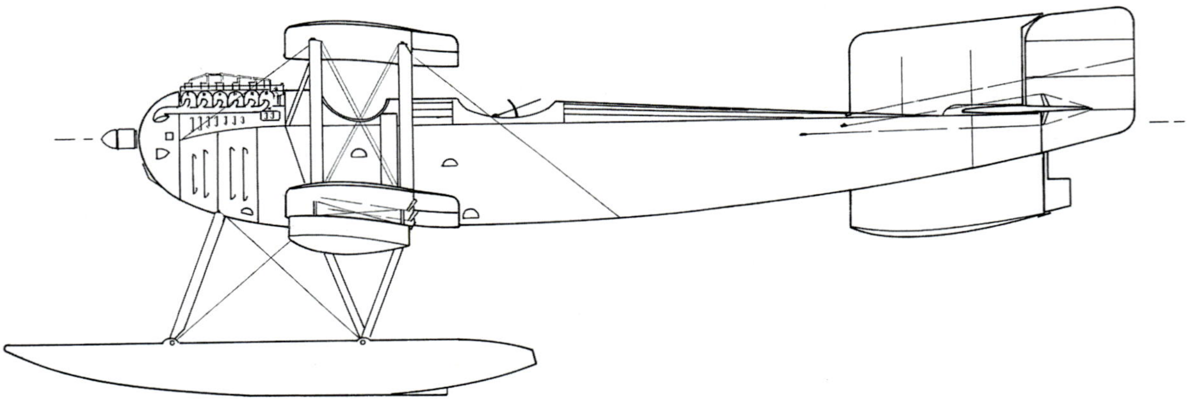

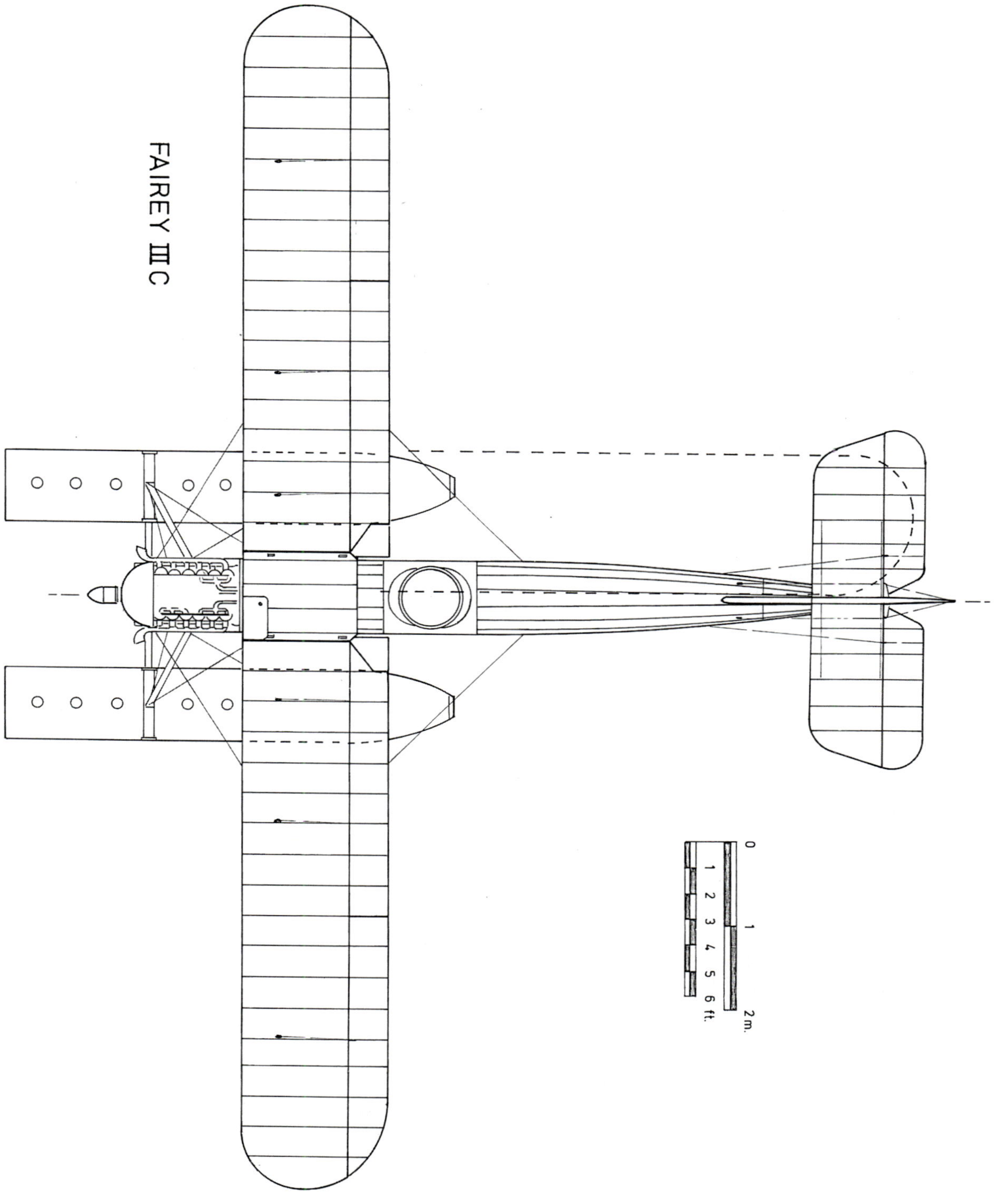
FAIREY IIIC
0 1 2 m.
1 2 3 4 5 6 ft.

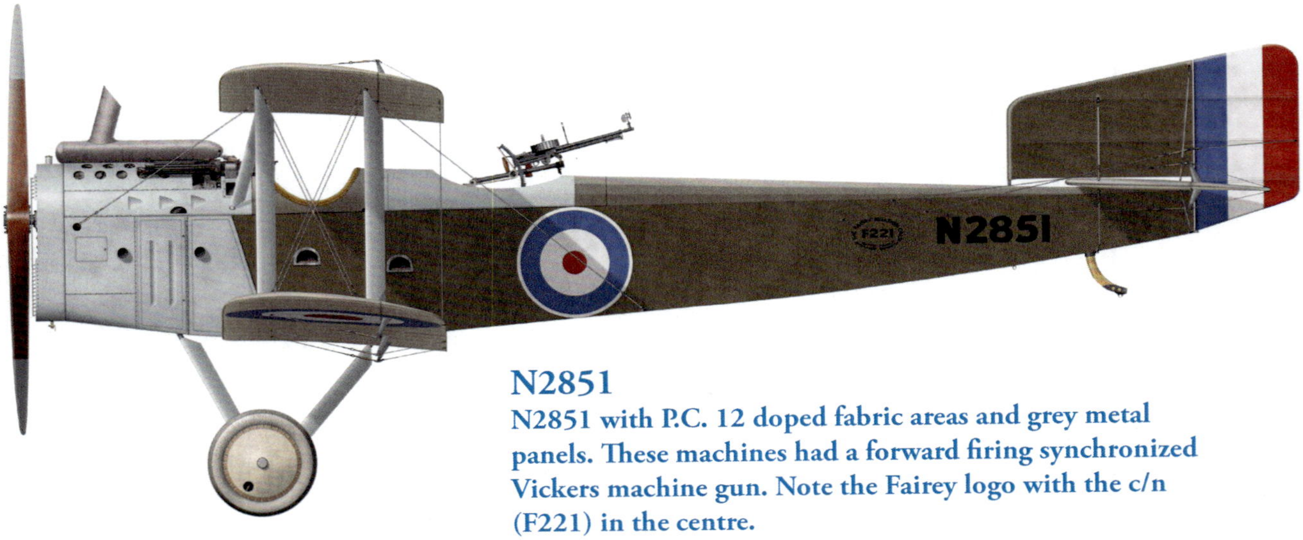

N2851

N2851 with P.C. 12 doped fabric areas and grey metal panels. These machines had a forward firing synchronized Vickers machine gun. Note the Fairey logo with the c/n (F221) in the centre.

IIIB

The serial of this IIIB is unknown. The floats were dark indicating a bitumen finish.

N2233

N2233 (c/n F280) in Soviet markings but still retaining the RAF serial but overpainted British national markings.

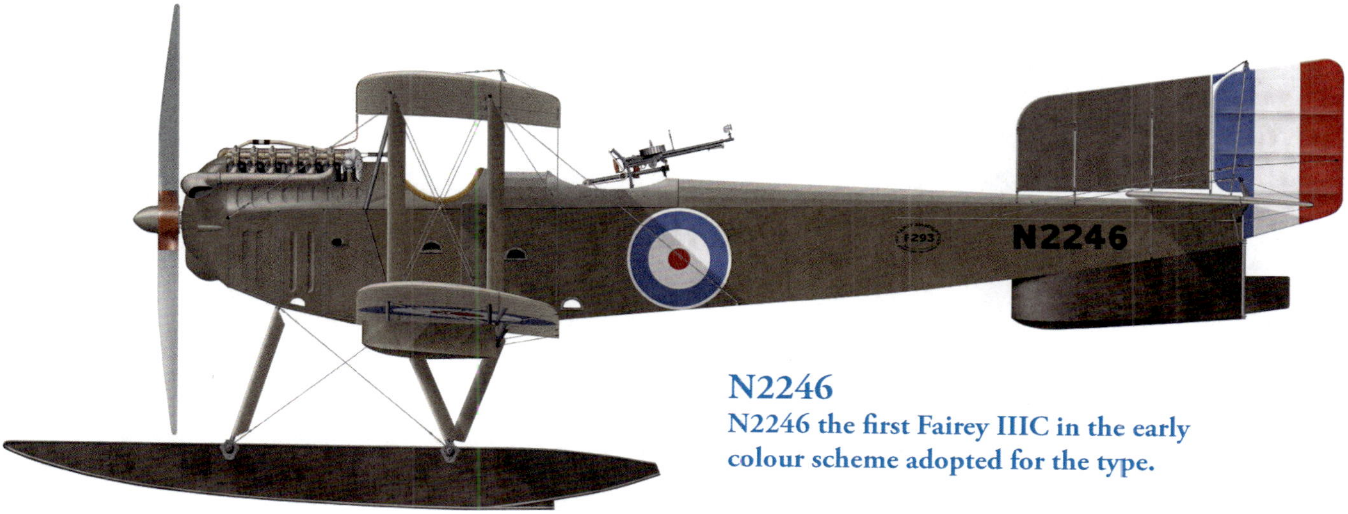

N2246

N2246 the first Fairey IIIC in the early colour scheme adopted for the type.

N9235

N9235 showing the application of the serial in white. On the central rudder stripe, it was in thin black outline. Metal panels area now battleship grey. Note the Fairey c/n F312 on the rear fuselage, and the Fairey logo on the wingtip float.

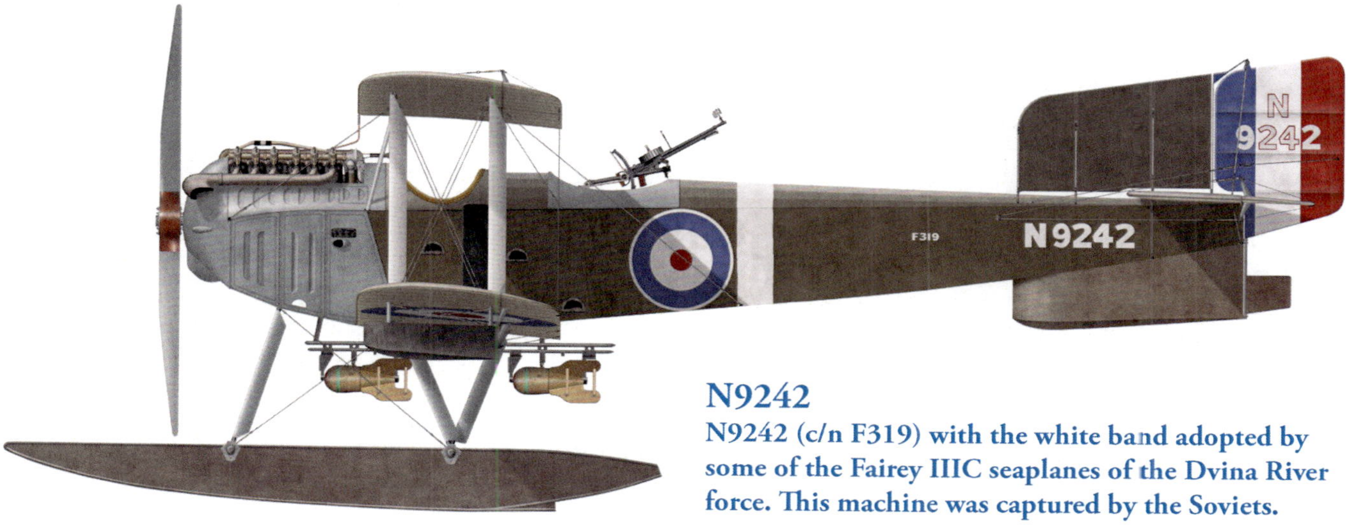

N9242

N9242 (c/n F319) with the white band adopted by some of the Fairey IIIC seaplanes of the Dvina River force. This machine was captured by the Soviets.

Chapter 7: Fairey N.2

At the 21st Admiralty Air Department Weekly Program Meeting in October 1917, it was stated that the present designs for ship aeroplanes had been got out *with the idea of using the machines only in seaplane carriers*. However, now proposals had arisen to carry single seater fighters in every light cruiser and battle cruiser and this meant that the Mann Egerton ship fighter was too large for the space available on these ships. It was said that the dimensions could not exceed those of the Beardmore W.B.III Folder development of the Sopwith Pup, and if possible, the new design should have the dimensions further reduced. For the present the Navy would have to rely on the Beardmore Folder until a new design could be got out by the Technical Department. It was suggested that the best line of development was to design a small machine around the A.B.C. engine.[1]

The A.B.C. engine had been used in the P.V.7 Port Victoria Kitten and the P.V.8 Eastchurch Kitten lightweight fighters. According to J.M. Bruce (Warplanes of the First World War – Fighters Vol.1) the Admiralty wanted to have numerous Zeppelin interceptors at sea and developed the Port Victoria designs to enable a torpedo-boat destroyer or similar ship to carry an aircraft.

The P.V.7 was built by the Experimental Construction Depot at Port Victoria. It had a wingspan of 18 ft, a length of 14 ft 11 in and weighing only 491 lbs loaded. The P.V.8's wingspan was 18 ft 11½ in, length 15 ft 7½ in and weighed 586 lbs loaded. The armament of both aircraft was a single Lewis Gun. The success of incendiary ammunition meant that this was a creditable armament for anti-Zeppelin operations. Construction of the P.V.7 began in the early spring of 1917.

Beardmore had commenced the design of their first totally original design, the W.B.IV shipboard fighter, around May 1917, before the above change in requirements. The W.B.IV and the contemporary W.B.V were large ship fighters with folding wings. The success of the Sopwith 2F.1 Ship's Camel fighter[2] apparently ended any hope of the Beardmore design being selected for production and it was abandoned before the end of the war.

The success of the Fairey N10 floatplane with the 260-hp Sunbeam engine led the Admiralty Air Department, in September 1917, to a proposal to fit it with a land undercarriage and conduct trials to see if it was suitable as a two-seater ship aeroplane. It was considered that there would be a demand for these machines in about six to eight months' time It was decided that there was enough time to order six experimental machines for trials before placing a large order if the type was successful.[3] These were ordered from Fairey and given serials N76 – N81. This proposed machine was a deal larger than the Sopwith 2F.1 and was a complete change from the smaller type proposed earlier.

At a later meeting, the *Fairey type N.2 a* machine was discussed. The loading of these machines was stated to be 8 lbs per ft^2 and the question was asked whether the landing speed would be low enough. It was pointed out that the machine was fitted with variable incidence gear by which means a low landing speed could be achieved. *Further, when actually flying with the Fleet the machine would usually only land when it had used most of its petrol, and therefore would be considerably lighter than when fully loaded.*[4]

The N.2.A specification was for a *Fleet reconnaissance work - flying off the deck Two seater with dropping land chassis and floating device.*[5] Previously the Type N.2.a was a seaplane or flying boat. The Fairey Campania had met the specifications for this earlier type.

Fairey were given an order for six tractor seaplanes to the Admiralty Air Department's Specification N.2(a) under Contract No. A.S. 30059 (BR.213). It seems that the Fairey IIIA was seen to meet the specifications and the contract was cancelled, as in January 1918, the Fairey N.2.A was discussed, *of which they had a contract for 50, and of which the first 10 were fitted or are being fitted with float and wheel carriages.* This would have been the batch of Fairey IIIA biplanes (N2864 – N2899) that was fitted with float, wheel and skid undercarriages.[6]

On 27 May 1918, Fairey wrote to the Director of Aeronautical Contracts on that date claiming that while no cost had been involved in the manufacture of the aircraft under the contract, *but we have a claim to make in connection with work done on Design.*[7] Lt Col A.J. Miley, the Assistant Controller, Design, wrote that it was considered that the firm should be refunded their costs in the design of the aircraft. The title to the minute was written in pen as "Fairey N10."[8]

Richard Fairey set forth a statement on the work done on the machines as follows: -

We were originally instructed by the Technical Department to design a Machine for use on board ship to a certain stated performance and with the capability of folding into an extremely small space. The machine was to be fitted with a 260 H.P. engine. The writer accordingly got out the design of the Machine, the drawing of which was submitted to Wing Commander Hewlett together with G.A. drawings of the Machine. The folding demands were satisfactorily attained by means of a new and special device and we were willing to guarantee that the machine should give the required performance. The writer, however, pointed out to Commander Hewlett the unsuitability of the engine for this particular design. Before submitting these drawings a great deal of calculation work was done, the Machine stressed throughout, the wing, fin, tail, upper area etc. Defined, a new form of chassis designed and all experimental drawings got out. In response to the writer's suggestion as to the unsuitability of the engine; it was arranged that only three of these machines were to be built and that the others should be equipped with the B.R. a rotary engine, and a new design was accordingly prepared to accommodate this. Owing to the material difference in weight and horse power of the engines, and almost completely new design was necessary. Work was in hand on this when the writer received verbal intimation not to proceed further as his services would be required on the design of the N.4 boat.

Fairey made mention that every machine that his company had designed had exceeded the specifications and complied with all other requirements. *We have, therefore, every reason to believe that the Machine would have passed its trials successfully.* He asked for the sum of £2,500 in compensation.[9]

Minutes indicated that the firm had spent about six months on the design work. The file ends without any indication whether Fairey received the asked for payment.

End Notes

1. Minutes of 21st Air Department Weekly Program Meeting 02.10.1917. TNA AIR1/270/15/226/113.
2. See Owers, C. *2F.1 Ship's Camel,* Datafile No.170, Albatros Productions, UK, 2015.
3. Minutes Royal Navy Air Department Progress Committee, 20th Meeting 25.09.1917. TNA AIR1/270/15/266/113.
4. Minutes Royal Navy Air Department Progress Committee, 21th Meeting 2.10.1917. TNA AIR1/270/15/266/113.
5. "Proposed Specification for Seaplanes and Ship Engines. Precis of Duties for Each Type." Attachment to Minutes Royal Navy Air Department Progress Committee, 25th Meeting of 30.10.1917. TNA AIR1/270/15/266/113.
6. Minutes Royal Navy Air Department Progress Committee, 33rd Meeting 22.01.1918. TNA AIR1/270/15/266/113.
7. Fairey to Director of Aeronautical Contracts, 27.05.1918. TNA AIR 2/969.
8. *Ibid.* Miley to C.A.5. 10.06.1918.
9. *Ibid.* C.R. Fairey, Sole Managing Director, The Fairey Aviation Co Ltd, to the Director of Aeronautical Contracts, 12.08.1918.

Fairey N10

Chapter 8: The Fairey Atlanta & Titania Flying Boats (Admiralty N.4 Flying Boat)

Above & Facing Page: Three views of the Atalanta under erection. The absence of any personnel in the photographs gives no clue to the immense size of the flying boat with a span of around 140 feet.

The Admiralty looked to building two types of experimental flying boats in 1917, and by January 1918 draft specifications for the Types N.3 and N.4 were prepared for discussion.[1] The types were taken over after the amalgamation of the RNAS and RFC as the RAF Type XXX and Type XXXIII respectively. The Armistice did not end the work on these flying boats, and they were considered important enough for their construction and testing to proceed even after the signing of the Peace Treaty that ended the war.

The USN was interested in the new flying boats that were in progress in 1918. The Felixstowe boats had proved their worth in the North Sea and the Curtiss Company was building the Felixstowe F.2a with Liberty engines as the H-16. In a Memorandum of November 1918, it was stated that there *is another radical flying boat to the N-4 specifications, designed by Major Linton Hope, which is not yet complete. The hull is complete and wings are being assembled at Litham* (Lytham?) *now. This boat is of similar construction to the two-engined P-5 only much larger. It is very spacious, having a gangway fore and aft. This boat is a marvellous piece of work embodying the most scientific truss ring construction, and sharp V-bottom and deformation of hull to prevent shock in landing. It is predicted that this boat will have a speed of 105 miles per hour.*[2]

This boat was the Fairey Atalanta.

The story of the Atalanta is not straightforward. According to H.A. Taylor (*Fairey Aircraft since 1915*, Putnam, UK, 1984) the Admiralty issued specification N.4 (later to become RAF specification XXXIII)[3] for a long-range fleet co-operation and open-sea reconnaissance flying boat. The object of this N.4 seaplane was to produce a machine that had the ability to rise from and alight safely in a sea way. It was initially proposed that it be powered by four 600-hp BHP engines, or four Rolls Royce Eagles if the BHP was not available in time, but at a reduced load.[4]

The specification was discussed at the Technical Committee meeting of Thursday 24 January 1918. The specification was sent to the Fairey Aviation Co on the 25th with a request that they look into the design.[5]

Above & Facing Page: The N.4 hull under construction at the Gosport Aircraft Co. (via P London)

A meeting took place between Lord Weir and Mr. Rutherford (representing Dick Kerr), Mr. Pybus (representative of the Phoenix Dynamo Co.), Mr Fairey of the Fairey Aviation Co., General Weir, Lt Col Ogilvie and Maj Miley from the Technical Dept.[6]

A recommendation for the design and construction of hulls for the N.4 seaplane was put forward in February that three hulls should be put in hand for the two seaplanes[7] being designed by Fairey Aviation Co and constructed at the works of Phoenix Dynamo Co and Dick Kerr.

First: Hull designed and constructed by Messrs Camper and Nicholson. This hull was required in order to give experience to commercial boat building firms into the design of flying boat hulls "in order eventually to dispense with the necessity of designing by Government Depts. or Stations."

Second: Designed by Lt Cdr Linton Hope and constructed under his direction at the works of May Harding and May. This firm was selected in order that the design and construction could be closely watched by Fairey and officers from the Technical Department. This firm's location was closer than Fife's works and in addition May Harding and May had experience in the construction of hulls along the lines of Linton Hope's proposals. This firm would be able to start construction much earlier than Fife that will only now be starting on this class of work.

Third: Designed by Lt Cdr Linton Hope and constructed at the works of Fife of Fairlie. Fife was considered in order that this firm would gain experience in the construction and design of flying boat hulls. Fife would send a representative to May Harding and May, to gain experience. Fairlie was too far away for the Department to keep a close watch on construction it was not considered practicable for Fife to build the only Linton Hope hull. Fife's construction would lag some weeks behind that of May Harding and May.[8]

It was considered that the experimental work at Felixstowe should continue *with a view of assisting private firms in the development of large boats, the experience with the new boat* (the Porte Fury triplane*) added to the previous experience obtained under actual service conditions being of great assistance to them.* This policy was seen as combining the best brains and all the financial and building facilities that was possible under present circumstances.[9] Authority to proceed was notified on 13 March.[10]

A Meeting was held on 17 April 1918, with regard to the completion of drawings and the construction of the two experimental seaplanes to the N.4 specification between officers of the Admiralty Technical Department and representatives of the firms concerned to discuss the best distribution of work in several drawing offices and shops with a view of completing the machines at the earliest possible date.

Present at this Meeting were Lt Col Ogilvie; Lt Col Hewett; Major A.J. Miley; Major P.A. Bumpus; Mr R.P. Wilson; Mr W.

Above: The N.4 hull at Gosport.

Leading Dimensions & Performance of N.4 Design by Mr. Fairey

Span Top Plane:	139 ft 6 in	Estimated performance	
Span bottom plane	92 ft 6 in	Full Speed	93 kts
Chord of Planes	12 ft 6 in	Getting off speed	48 kts
Gap of Planes	13 ft 0 in	Climb from sea level	850 ft/min
Total Lifting Surface	29,000 sq ft	Service Ceiling	15,000 ft
Overall height	29 ft 9 in	Load to be Carried	
Overall length	66 ft 9 in	Crew of 6	
		Fuel for 6½ hrs at 90 kts at sea level.	
Minimum estimated weight	30,370 lbs	One Coventry Gun and 50 rounds of ammunition.	
Maximum estimated weight	31,770 lbs	4 Lewis Guns and ammunition.	
		Long range W/T.	
		Bombs	2/500-lbs.

Manning (Phoenix Dynamo Co Ltd); Mr C.R. Fairey (Fairey Aviation Co Ltd); Mr V. Watlington (Dick Kerr & Co Ltd); Mr P.J. Pybus (Phoenix Dynamo Co Ltd); Capt J.C. Highfield (Phoenix Dynamo Co Ltd); Mr W.E. Highfield; (Dick Kerr & Co Ltd); and Mr C.E. Nicolson (Camper & Nicholson Ltd - Gosport Aviation Co).

It was desired to get the N.4 type experimental flying boats up to production stage at the earliest possible moment. Fairey would have the general design and layout of the job and would be assisted by the Phoenix and Dick Kerr in getting it into production. Problems were foreseen due to the Man Power Act that claimed men under 26 years of age for military service. This was thought to impact on the draughtsmen that were required for the project.

The problem with having drawings made at the different locations made standard was discussed. It was agreed that the separate firms would send photographs of the drawings every day to Faireys who would coordinate the project. Mr Pybus said that he would guarantee that no delay would result from each firm constructing one machine.

It was agreed that in order not to limit the production capacities of the boats owing to the difficulty of road transport the design should include folding extensions and kingposts as well as making the lower plane centre section detachable from

Above & Below: Two views of the hull at Gosport. (Gosport Aviation Society via P London)

Above: Moving the N.4 hull was a massive logistical exercise. (Gosport Aviation Society via P London)

the hull. At this time all three firms had not received an official order.[12]

A requisition was submitted on 2 April and approved on the 14th. An order to proceed was then sent to Phoenix on 9 May, and to Dick Kerr on 10th.[13] Fairey wrote to the Technical Dept in May stating that he had not received a contract although they had already had a considerable amount of work in hand. As Mr Nicol, Director of Contacts, had informed Fairey that he should undertake no work without a contract as he had pointed out on a previous occasion that he was unable to pay for some items that were not listed in a contract. *The experimental work on which we have been engaged to date, although our machines have succeeded in the Technical sense, has involved us in heavy loss, for which we have had no return and we do not think we should be asked to continue to work at high pressure until definite terms of payment have been agreed and a proper Contract entered into. Moreover, as we hope shortly to be in a position to issue the first batch of drawings, we consider a definite understanding should be arrived at before any of these are passed onto Messrs. the Phoenix Dynamo Company with whom we have no arrangement whatever.*[14] This letter drew a quick response. *It had been agreed that this firm will do the design of this machine and Messrs. Dick Kerr and Messrs. Phoenix will actually manufacture it.* The Requisition only covered the latter two firms and it was requested that Fairey's position be regularised by the placing of a definite order for the work that *is now well advanced.*[15] Another letter noted that *from his letter that his energies, which should all be devoted to the urgent question of getting out the design for the big flying boat, are being, from our point of view, wasted in worrying over financial matters.* It was noted that if Fairey had not received any payment for designs which had been accepted for production, then he would have a reasonable cause for grievance.[16] This was to continue for a long time as will be related.

At the same time Porte was developing the P.S.B. (Porte Super Baby later to be christened the Fury) triplane flying boat with five Rolls Royce Eagle VIII engines. *The machine is an intermediate step between A.F. Specifications N.3 & N.4 and is being built with a view to gain experience in the design of larger seaplanes capable of landing and rising from the open sea under average North Sea conditions.*[17] In order to get the initial experiment carried out as quickly as possible Porte was pressing ahead with the design of a machine fitted with five Rolls Royce Eagle VIII in order that the experiment should not be delayed waiting for the 600-hp BHP or Rolls Royce engines.[18]

Fairey was to design the aerostructure with the hulls being to a Linton Hope design, Fairey having purchased the rights to Linton Hope's design. Fairey was to have overall responsibility for the project.[19] Hulls were to be built by Fife[20] in Scotland and May, Harding and May at Hayes.[21]

To complicate the story a further contract was let to the Gosport Aviation Co for a hull to that firm's design.

Above: N119, the Atalanta, resplendent with varnished upper hull and white anti-fouling paint to the bottom of the hull, aluminum doped fabric surfaces.

Above: Another view of N119 after erection. (N Cooper via SEAWINGS)

Above: Another view of N119 after erection. (N Cooper via SEAWINGS)

It has not proved possible to confirm all the serial numbers assigned to the different boats as will be explained hereunder.

The progress of the project is recorded in the Technical Committee's fortnightly reports as follows:

Fairey Atalanta. Contract for design of aerostructure only.[22]

F/E 12.06.18.

Work on the drawings is proceeding satisfactorily. Stress calculations on the engine mountings and centre section are in hand.

F/E 26.06.18.

The procedure as to the issue of drawings has been laid down and a representative of Messrs The Fairey Aviation Co, has gone to Messrs Dick Kerr with the first instalment of drawings.

F/E 07.08.18.

Much better progress on drawings is now being made; the first instalment of drawings of the Centre Section has been sent to Messrs Dick Kerr. A mock-up of the hull is well advanced and steps have been taken to improve the pilot's view. Further draughtsmen have been engaged, drawing work on the tail unit has been possible.

Structure of main planes has been approved for strength.

F/E 21.08.18.

Work is proceeding satisfactorily. Mock-ups of the C.O. (*sic*) gun and wireless gear have been prepared and added to the hull mock-up. A mock-up of the top plane fighting positions is in preparation.

F/E 04.09.18.

Good progress has been made with the drawings, the centre section structure being practically complete.

The firm has been instructed to proceed on the assumption that Condor engines will be fitted.

F/E18.09.18.

Design work is well ahead of manufacture. Visits have been paid to the factory firms to hasten construction. The method of fastening and fairing the bottom centre plane to the hull has been one into and mocked up.

F/E 02.10.18.

Drawing Office work is proceeding satisfactorily, and is now concentrated on Tail Unit, Engine Mounting and Hull fixtures.

F/E 16.10.18.

Drawings are making good progress.

F/E 30.10.18.

First time called Atalanta.

Good progress has been made in Drawing Office. Self-sealing tanks have been arranged for.

F/E 13.11.18.

Work is progressing with tanks and petrol systems, also adjusting tail gear, and further details of internal fittings to hull.

Close co-ordination with Messrs Dick Kerr is being secured.

Above: This view of N119 with personnel working around the machine shows the huge size of the seaplane.

Above & Facing Page: The N.4 in its natural elements. (P London Collection)

N119

N119

Above: A post card that was altered to make the Atalanta appear to be flying.

F/E 27.11.18.
Work is advancing.

F/E 11.12.18.
Work on drawings of this machine is making steady progress.
Period 12 to 31.12.18 (inclusive).
Work is proceeding satisfactorily.

Month ending 31.01.19.
Seating arrangements have been decided on. Tank drawings are issued. Check sand tests to determine importance of glue joints have been carried out.[23]

Fairey/Phoenix Atalanta N118.[24] **Contract for construction by Phoenix of one machine to Fairey design.**

F/E 12.06.18
Construction of Fairey Aerostructure. Construction of one machine. Work not yet commenced.

F/E 26.06.18.
All drawings for this machine will be issued from Messrs Dick Kerr's Drawing Office.

F/E 07.08.18.
Drawings still being checked and adapted for production. Construction not yet started.

F/E 21.08.18.
No work yet commenced on this machine.

F/E 04.09.18.
Work not yet commenced. A few drawings have just been received by Messrs Phoenix and they have promised to get them into the shop without delay.

F/E 18.09.18.
Drawings are only just beginning to come through from Messrs Dick Kerr so that little actual work has so far been possible.

F/E 02.10.18.
N118. A few more drawings have been received and templates made. A few wiring plates in progress.

F/E 16.10.18.
Machine N118. Spars for centre section in progress. Also compression struts and end rib stiffeners. Further progress has been made with templates, wiring plates, and wing fittings. Drawings are coming through faster.

F/E 30.10.18.
First time called Atalanta.
N118. Progress is fair and in general up to the output of drawings and spars, drift sturts, ribs and main fittings are in hand.

F/E 13.11.18.
N118. Components are well in hand in accordance with drawings so far received.

F/E 27.11.18.
N118. Work is well advanced in accordance with drawings received.
Report by Maj Hume, RAF, on visit to Phoenix Co on 04.12.18.[25]
The firm is delayed on this machine for want of a man from Messrs. Fairey Aviation Co. to explain certain pints not clear from the drawings. The firm have written to Mr. Fairey about this but seem to be receiving poor support compared with Mr. Dick Kerr.

Above: N119 ploughing through the water, probably off Felixstowe. Note that even with the engines mounted high a lot of water appears to be thrown into the propellers arc.

The question of transport was raised...Lytham is regarded as very unsuitable and the firm would like delivery by road please. The extra height occasioned by the permanent stump stern post was brought to my notice; this will probably mean a modification in the present transport trolley.

F/E 11.12.18.

N.118. Work has been slightly delayed pending a visit from Messrs Fairey to elucidate certain drawings. This has now been done and work is proceeding in accordance with drawings received.

Period 12 to 31.12.18 (inclusive).

N118. Work proceeding satisfactorily in accordance with drawings received

Month ending 31.01.19.

N118. Spars are well in hand. Trouble has been found with packing blocks splitting. Steel fittings are in hand in accordance with drawings received. Ribs glued in a cool shop have been found unsatisfactory, have been replaced. Progress is slow.

It is believed that N118 was completely finished and used for flotation tests but never flown.

Fairey/Dick Kerr Atalanta N119. Contract for construction by Dick Kerr of one machine to Fairey's design.

F/E 12.06.18

Work not yet commenced.

F/E 26.06.18.

The first instalment of drawings has been received from Messrs Fairey. These include all main plane details. These drawings are being checked and re-issue to the shops and to the Phoenix Dynamo Co.

F/E 21.08.18.

Drawings are still being co-ordinated by Messrs Fairey's resident representative. The construction of component parts has commenced.

F/E 04.09.18.

About one-third of the number of drawings expected to complete this machine have now been received from Messrs Dick Kerr.

18.09.18.

Hastening action has been taken with this firm and the extreme importance of this experimental design fully brought home to them. A few small fittings, jigs and templates only have so far been made. The firm has undertaken to provide special arrangements for the construction of their first experimental machine in order to hasten completion.

F/E 02.10.18.

N119. Manufacture of detail parts proceeding on Main Planes in Templates for fittings.

Above: N119 on the slipway at Felixstowe.

F/E 16.10.18.

N119. Progress is still slow. Timber is cut for a good portion of the job but not yet worked up. A number of wing fitting plates have been cut out. The question of a suitable trolley, which through a misunderstanding has been in abeyance, has been taken up with the firm and a draft design is expected shortly.

F/E 30.10.18.

First time called Atalanta.

N119. Progress is still lamentably slow. It is reported that less than 20 men are engaged on this machine. Spars, ribs and fittings are reported to be in progress. The firm will receive their hull next week.

F/E 13.11.18.

N119. The work is still slow on this contract, but strong hastening action has been taken. Main planes are now being assembled.

F/E 27.11.18.

N.119. Considerable improvement in progress is now apparent, several sections of the main planes are being assembled and fittings are well advanced.

F/E 11.12.18.

N119. Main planes and centre section are being assembled. Spars for wing roots are in hand and metal fittings are progressing.

Period 12 to 31.12.18 (inclusive).

N.119. Main planes and centre section assembling. Metal fittings well in hand.

Month ending 31.01.19.

N119. Wings assembled and several fittings completed. Weights appear to be below estimates. Tanks in hand. Progress slow. Nothing done to hull.

In his Report for the 16th Maj Bumpus noted that while progress "on the whole is still slow but the work done is good."[26]

According to Ron Smith, the hull was built by May, Harden & May and taken by road to Dick Kerr's works at Lytham,[27] where, on 22 October 1919, the Deputy Director of Design, Wing Cdr Briggs, together with Flt Lt Maycock of the D of O, and Mr Fairey made an inspection of the N.4 boat. The wings and tailplanes were in position but had not been covered. The controls were connected up. Fairey stated he was satisfied with the work so far completed. Allowing for some delay in the supply of Condor engines it was expected that trials could be carried out in spring the following year. Some of the problems of operating these large flying boats are indicated in the requirements that had to be addressed in order to test the boat.

Before the flying boat is launched it will be necessary to lay three 10 cwt. Moorings with light buoys attached in the following positions –

Above & Facing Page: The second N.4, the N129 *Titania*. The major difference between the two flying boats was the Fairey patent camber gear.

One each side of the slipway about 150 ft. distant from the centre line opposite the position where the slipway descends steeply towards a gulley due south.

A further 10 cwt. Mooring with life buoy attached on the mud at the far side of the gulley.

These moorings and buoys will be required to correctly align the hull on the trolley should it be necessary to withdraw the flying boat on the water. For the purposes of the trial it will be necessary to lay two moorings in the neighbourhood of Lytham pier, clear of the navigation fairway. ... Owing to the large rise and fall of the tide it is not considered practicable to conduct the trials unless these latter mooring out positions be supplied.

Messers. Dick Kerr & Co. stated that they are prepared to supply a heavy lorry to assist in drawing the trolley up the slipway but it is considered that a handling party of approximately 50 men, with suitable percentage of waders, would be required. Owing to the depth of water drawn by the hull when floating difficulty will probably be experienced in accurately adjusting the trolley under the hull but it is hoped that the trolley now being constructed at Preston will suffice for the trials and the experience gained will decide whether a new design will be essential under ordinary working conditions. [28]

Linton Hope/Fife. Contract for construction of one hull to Linton Hope lines and Fife details.

F/E 12.06.18
A number of drawings have been sent to the firm and construction is proceeding.

F/E 26.06.18.
Construction is proceeding satisfactorily.

F/E 07.08.18.
Work on this hull has been partly delayed owing to the loss of four boat-builders, who have been requisitioned for repair work on Stations, also some joiners who have been called up for the Army.

F/E 21.08.18.
Some difficulty has been encountered in bending the stringers round the bow. Progress has been poor and hastening action has been taken.

F/E 04.09.18.
The difficulty experienced in bending the stringers round the bow of this hull have been overcome and the stringers are now in position. It is expected that timbering will commence this week.

F/E 18.09.18.
Work is proceeding satisfactorily and the hull is now being timbered.

F/E 02.10.18.
Timbering on this hull is finished and two-thirds of first skin is in place. Wing Tip Float Moulds partly made.

F/E 16.10.18.
The hull is nearly complete. Stay plates are on and the planning bottom well advanced. It is anticipated that varnishing will start this week. Wing floats are well advanced and the bottom planked. Pilot's cockpit fairing equally well advanced.

F/E 30.10.18.
The hull is progressing well; it is fully planked and the planing bottom is being prepared. Wing root stay plates are well in hand. The wing float keels are aid.

F/E 13.11.18.
Steps are now being framed up and planked, so that the hull is nearing completion.

F/E 27.11.18.
The hull is well advanced and nearing completion without further difficulties. The wing floats are progressing well.

F/E 11.12.18.
This hull is now having the forward planning bottom framed up. Steel fittings have gone away for sheradizing. This hull will be available at the New Year. One wing float is nearly done and the other is in frame.

Period 12 to 31.12.18 (inclusive).
Approaching completion, waiting return of wing root fittings from sherardising. Floats are well advanced.

Month ending 31.01.19.
Ready for delivery.

<u>Linton Hope/May Harding & May.</u> Contract for construction of one hull to Linton Hope lines and details.

F/E 12.06.18
Good progress is being made. All stringers are in position and nearly all the inner hoops and all the spar straps.

F/E 26.06.18.
Great progress has been made on the hull. All timbers are in position and planking is well in hand. A model has been run in the William Froude Tank at the N.P.L., and modifications are being introduced to obtain the best possible results. Another model with a slightly modified fin will be run this week.

F/E 07.08.18.
This hull is nearly finished and is having steps fitted. A complete set if construction photographs have been made for record purposes and for information of subsequent boat-builders.

F/E 21.08.18.
The hull is nearly finished, but the planning bottom cannot be completed pending the delivery of the wing root stay fittings, which are well in hand. The A.I.D. have inspected this hull for material workmanship and given a highly

Above: British America Tobacco Co cigarette card - The Fairey *Titania"* The largest flying-boat in the service of the British Royal Air Force, The *Titania* has been designed to operate for long periods of time away from its base, and full sleeping accommodation provided for tis crew of six. The span is 139 feet, and fully loaded it weighs almost 15 tons. Four Rolls-Royce "Condor" engines carried between the planes, drive the air-screws, two in front of the planes and two to the rear of the planes. In the hull is the pilot's compartment, in front of which sit the gunner and observer. Because of its size the N.4 boats were favourites for Cigarette Cards in the 1920s..

favourable report.

F/E 04.09.18.

Work has been somewhat delayed pending delivery of the Wing Root Stay fittings. It is expected that these will be completed and fitted shortly, when about another ten days will finish this hull.

F/E 18.04.18.

This hull is half turned over and awaiting wing root stay plates and other fittings which are made and are now being annealed. A pilot's hood has been designed and is being fitted and Mr Fairey has been in consultation on extra internal stringers to take seats, etc. Bituminising is complete and varnishing is in hand. The wing tip float keels are laid.

F/E 02.10.18.

Annealed metal fittings received, and the forward one is in place. Moulds are up for the hood. This will probably be the hull for N119. Wing Tip Floats are well advanced and planking is about to commence.

F/E 16.10.18.

No report.

F/E 30.10.18.

The hull is now finished and will be delivered to Lytham on October 29th. It weighs 2,982 lb bare or ½% over estimate; this with addenda of fairings and fittings will be 10% of total weight of machine. The hull is an excellent job in every respect. Wing floats are well advanced.

F/E 13.11.18.

Despatched to Messrs Dick Kerr October 30th. Wing tip floats proceeding satisfactorily.

F/E 27.11.18.

Hull delivered. Wing tip floats nearly ready.

F/E 11.12.18.

Hull delivered. One wing float is complete and being varnished and the other is being planked.

Period 12 to 31.12.18 (inclusive).

The hull is delivered. The wing floats are now complete and ready for despatch.

Month ending 31.01.19.

Delivered.

Although not stated, it is presumed from the above reports that the finished hull and aerostructure would have been shipped to Fairey for final assembly and fitting out.

On 6 November 1918, Fairey had been notified that it had *been decided that the first experimental machine of the* ...(N.4)... *class shall be fitted with ordinary steel tanks but that production drawings shall be produced with to suit self sealing tanks which, should the type go to production, will be the approved type.*

This procedure is designed to save time as the vulcanising plants are very fully occupied upon production work.[29] Fairey was to notify

Above: This is believed to be the hull of the Atalanta when used as a houseboat in the 1950s.

the contractors of these facts.

A USN report with respect to the Phoenix P-5 also contained reference to the N-4 and while it would be logical to assume that it refers to the N-4 hull built by Phoenix, a later, but almost the same, report states that it was inspected in the May, Harding and May yard.

A new hull to be known as the N-4 is being laid down. Its dimensions will be:

Length . . 64.0 Ft
Beam ..... 8.5 ft
Depth .. . 9.0 ft.

The displacement of the completed boat is 33,000 lbs.

In this hull an attempt is made to reduce the amount of added structure for the planes to a minimum. Except for a short way forward of the step, the main plane is formed by a simple doubling of the double planking – making four thicknesses – and the flaring edge support is the only structure.

This flaring edge seems to be highly important.

Tank runs show its effect in assisting the "getaway" and in suppressing "Blisters" which might rise into the propellers is very marked.

The proportional weight saving becomes more apparent in this size hull. The longitudinal are built up of five pieces for instance, instead of being solid, and may be reduced at the ends as found desirable. The estimated weight given is thought to be safe. For a craft having a flying weight of 33,000 lbs the hull weight of 3,300 lbs seems very good.

The wing spars instead of entering the hull will rest on top of it in this design. The hulls were "seen in the maker's yard on April 13th."[30]

A further USN report of 28 November 1913, states that there *is another radical flying boat of N-4 specifications designed by Major Linton Hope, which is not yet completed. The Hull is completed and wings are being assembled at Litham* (Lytham?) *now. This boat hull is of similar construction to the two-engined P-5 flying boat only much larger. It is very spacious, having a gangway fore and aft. This boat is a marvellous piece of work embodying the most scientific truss ring construction, circular in cross section with beautiful streamline shape, and sharp v-bottom and deformation of hull to prevent shock in landing. It is to be a biplane with installation of four 600 H.P. motors in tandems. It is predicted that this boat will have a speed of 105 miles per hour....*

The report goes on to compare the progress in large flying

boats in comparison with landplanes. *The new circular boat hull constructed in N-3 and N-4 Specification in comparison to the Super Handley Page and Tarrant Triplane land machines shows that for a flying machine of the same strength the boat hull is a smaller per cent of the total weight than the fuselage of the land machine. In addition to this we have the added necessary weight of the landing chassis in a land machine which is not true of flying boats. This is a very interesting discovery, and such a thing which otherwise would have washed out large flying boats in favour of land machines for flying over either land or water.*[31]

The hull construction of the Linton-Hope N and P types of boats comprised a number of light stringers around the more or less circular sections of the boat. Ten men had built the May, Harding and May hull in 13 weeks – three being experienced boat-builders, five joiners and two apprentices. While it took seven men – two experienced boat-builders, the rest the same, 11 to 11½ weeks to construct an F-5 hull, the rate of planking of the N.4 was greater giving a better utilisation of man power according to David Nicolson.[32] The N.4 hull was estimated to be 2,970 lbs and was actually 2,982 lbs, a variation of less than 0.5%. "The position of the centre of gravity in this case was exactly as designed."[33]

Wood was bent into a circular hoop shape to the cross section of the hull. The timbers were all in one piece bent around the hull with the ends at the keel into which they were jogged and glued to the longitudinal stringers. The planking was applied over this. Narrow Honduras mahogany strips were laid diagonally to form the inner skin. The same wood was used for the outer skin but laid fore and aft. Varnished nainsook was laid between the skins. This was not only to make it watertight but it is light and very strong. The inner skin was held in place by copper nails. The planning bottom was then attached to the hull.

The N.4, which was 2½ times the bulk of the F.5, the hull design weight was 2,970 lbs with no allowance for unforeseen weights, and the actual weight was 2,982, an error of 1 lb or under ½%. LH – Major Linton Hope - wrote in his report that "This form of construction therefore gives greatly increased strength for weight, and very accurate weights to estimates."[34]

The first Atalanta to be completed, N119, utilised a hull built by May, Harding and May, and received the Fairey construction number F.276. This was the Dick Kerr machine. Despite being almost completed by the Armistice this machine did not fly until 1923.

The Fairey Atalanta was the largest flying boat in the world at the time of with a span of about 140 feet and a gross weight of more than 30,000 lbs. The four 650-hp Rolls Royce vee-12 liquid-cooled Condor 1A engines were arranged as tandem pairs. The three-bay biplane had unequal span wings with the top wing overhang supported by kingposts on the upper wing. The tailplane was a biplane structure with three fins and rudders. The open cockpit provided seating for two pilots with a gunner's position forward. Two oval cut-outs in the rear hull allowed for beam guns.

The hull for N119 was dispatched from May, Harding and May at Hythe, Southampton, to Dick Kerr at Lytham on 30 October 1918, for the fitting of the aerostructure and internal fit out.[35] N119 apparently remained at Lytham in an assembled but unfinished state until 1921.[36] It was disassembled and taken by road to the Isle of Grain where the flying boat experimental station was situated at the time.

N119 underwent taxying trials at Grain in 1922. The seaplane was taxied at about 10 knots from her moorings into the Thames estuary where there was practically an unlimited run. Two motor boats were employed to run at full speed into the wind, one about a quarter mile astern of the other. Each boat carried a photographer and two observers. The trial run was arranged such that the seaplane would pass by each boat in succession.

It was arranged for the second pilot to fire a flare at various times noting the speed when he did so. An observer in the seaplane was to note the engine rpm at the same time.

The machine reached a maximum air speed of43 knots, but was not then running on the step. It was estimated that the "dirtiest" speed occurred at about 15–20 knots.

A second run could not be carried out due to the damage caused to the aft starboard propeller by spray. The fabric covering the forward starboard propeller was damaged such that it had to be removed before the seaplane could proceed to her moorings under her own power. The fabric covering the port propeller had been removed previously due to damage caused from a preliminary handling trial on the water. It was stated that at no time were the engines at full throttle but that all throttles were opened together by means of the master control that gave very nearly equal openings.[37]

Principle difficulties found with the design were that *the water thrown up by the hull breaks the propellers and failure of the servo-motor gear for operating the tailplane. The next attempt to fly this machine should be made towards the end of November* (1922).[38]

It made its first recorded flight on 4 July 1923. It then went to Felixstowe, the Marine Aircraft Experimental Establishment (MAEE) having moved there early in 1924.

In May 1924 *Flight* noted that the Atalanta with its four Rolls-Royce Condor engines of 700-hp each was *probably the largest and most powerful flying boat in the world, and has recently passed its tests successfully. Beyond stating that the "Atalanta" is a six-seater sea reconnaissance flying boat, no particulars may be published.*[39]

In May 1921 it had been decided to conduct experiments regarding lateral stability and control of aeroplanes with models of three different types of machines: the S.E.5a, Bristol Fighter and Fairey N.4 flying boat. At the time it was thought that sufficient full-scale work could be carried out to enable a comparison with the model data. However, the large amount of work involved in trying to obtain quantitative data on a large flying boat led to the N.4 research being limited to the investigation of lateral stability only. A 1/30th scale model of the Fairey N.4 Atalanta flying boat was tested in the 7 ft wind tunnel No.2 at various angles of yaw and a number of reports issued.[40]

The second seaplane to be completed was N129 (Fairey c/n F.337), the Phoenix built machine. The other machine apparently had the serial N118. Why there is such a gap between the serials is unknown however a Note accompanying a letter of 12 June 1919, states that the Fairey Atalanta N.4 was *now being erected at the Phoenix Dynamo Co.*[41] According to Sturtivant and Page the contract for N129 was issued post-war. N129 was given the name Fairey Titania.[42]

The principle differences between N119 and N129 were the mark of engine and that the Titania had the Fairey patented camber-changing gear. A Report of 5 November 1919, stated that the proposal by Fairey to fold the wing extensions was agreed to and would be incorporated in the Titania only. It was proposed to hinge the top plane extension and the lower plane extensions to fold upwards into line with the interplane struts. This would allow for an increase in weight of about 30 lbs but it would reduce the stowing dimensions to 79 ft wide and 70 ft long (without COW gun), and reduce the height by about 2 ft. *This will allow the existing Hamble sheds to be used for erection.*[43] N129 had Condor III engine of similar power and was slated for an all up weight of 30,500 lbs. Both had an endurance of 7 hours with a max of 9 hours, could carry a 1,000 lb bomb load and had provision for up to six Lewis guns. This would imply that all positions had a double Lewis mounting.

The hull of N129 was constructed by Fyffes on the Clyde in Scotland. This company is spelt Fife in the Technical Dept Reports. The hull was sent to Fairey at Hamble by road. The movement of these massive hulls was a major undertaking and special routes had to be reconnoitred to allow for their size avoiding such obstructions as trams and narrow streets. Bruce Robertson wrote that the Atalanta's *hull probably travelled more miles by road than it did in the air.*[44]

The aerostructure was sent from Hayes to Hamble where the boat was assembled. *Flight* for 10 March 1921, reported that the Titania was now finally being erected with flights expected to take place within the next month or six weeks. *The Times* stated that the *work for which this giant aircraft has been designed and built is long distance sea patrol. She has a range of 1,500 miles and will be capable of remaining at sea for some days at a stretch. She will operate from a sea base much in the same way a submarine returning to port for stores and fuel. The Fairey flying boat will be able to dispense with anything in the nature of a parent ship. She will carry a crew of seven, and has emplacements for five machine guns, of which a proportion are fitted in the wings.*[45] Although these reports reported that the machine would fly soon, this was not to be the case.

The hull had to be sent to Hayes for modification and then returned to Hamble where it remained until 1923.[46] The machine was disassembled and transported to Grain for reassembly and testing. The first flight appears to have been made in 1925 from Felixstowe where the Marine Aircraft Experimental Establishment had moved to in June 1924. The machine was in service from here until 1928 or 1929.

Gosport Aviation Co. Contract for the design and construction of one hull.

This third machine was designed by Charles Nicholson of Camper and Nicholson, designers and builders of yachts.[47] The Gosport Aircraft Company of Gosport, Hampshire, was formed at the start of the First World War by Sir Charles Allom of White, Allom and Co, and Camper and Nicholson, hence Charles Nicholson's involvement.[48] The company built a number of flying-boats for the British Government and Charles Nicolson was a renowned designer of yachts which explains why they were selected to build a hull to the N.4 specification.

The difficulties of understanding the development of these machines is illustrated by the data in Ray Sturtivant and Gordon Page's magnificent work *Royal Navy Aircraft and Units,* (Air Britain, UK, 1992). According to them the following contracts were issued for the Atalanta flying boats:

Contract No. 38a/374/C.359 of May 1919, was raised to cover one Fairey Atalanta I reconnaissance tractor flying boat from the Gosport Aviation Co. The serial N118 and Fairey c/n F.275 is given for this aircraft that was built by the Gosport Aviation Co. As noted above N118 was the aircraft built by Phoenix.[49]

Contract No. 38a/373/C.360 of 13 May 1919, is given to the Nicholson (*sic*) hull built by May, Harding and May. To this aircraft they allocate the serial N119 (Fairey c/n F.276). However, N119 was the Dick Kerr machine. There is no other documentation that Nicolson had any input into the May, Harding and May hull, in fact the reports indicate that the Gosport machine was considered separately to the other two.

The progress of the Gosport design is recorded in the Technical Committee's fortnightly reports as follows:

F/E 12.06.18

Two models of this hull are now at the N.P.L. for tests in the wind tunnel and in the Froude National Tank which will e carried out this week. This hull both in design and construction is entirely original and therefore construction is therefore delayed pending the N.P.L. report.

F/E 26.06.18.

A model of this hull was tested at the William Froude Tank at the N.P.L. The trial showed that the water resistance of the hull is too high and that the pitching moment required to prevent porpoising is impossible of attainment in the full-sized machine. The firm are modifying the design accordingly.

F/E 07.08.18.

Web and bulkheads in good state of progress and all materials in preparation. The firm expect to complete the hull early in November.

F/E 21.08.18.

This hull is now up in frame and a more detailed report and inspection will be made next week.

F/E 04.09.18.

Frames of this boat have been erected and stringers fitted, but the latter have been taken out again for alteration. The hull should be completed by the middle of October.

F/E 18.09.18.

Progress and workmanship are satisfactory. Stringers are all in and the fin chines are fitted. Planking should start immediately.

F/E 02.10.18.

Forward gun coamings area fixed and structure is being faired up preparatory to planking which has now commenced. Wing Tip Floats are progressing.

F/E 16.10.18.

Planking is proceeding on this hull and general progress is satisfactory. Drawings are in hand for the Wing Root attachments.

F/E 30.10.18.

The first skin of the planking is complete, and the outer skin begun. This hull has been promised for the middle of November.

F/E 13.11.18.

Illness has slowed up this work, but it is now progressing well enough. Limbers to top sides of floats are finished. The hull is now very nearly completely planked, and is drawing near completion.

F/E 27.11.18.

The final planking of the wing root stay plates is still under discussion. Wing floats are progressing well.

F/E 11.12.18.

The planking of the bottom is nearly complete, The wing root fittings are still under consideration.

Period 12 to 31.12.18 (inclusive).

Approaching completion. Fore Hatch hood in hand; wing floats being decked in. Waiting for drain plugs.

The Report of Maj Bumpus of 1 January 1919, noted that a *Sketch design has been prepared of an internal strengthening for the wing root tube attachment of the Nicholson Hull and drawings of the hull have been requested from Mr. Nicholson to enable progress to be made on the internal arrangements of the hull.*

Mr Nicholson is to visit Messrs Fairey during the next few days to discuss this and other outstanding points.[50]

Month ending 31.01.19.

Awaiting wing root attachment fittings which have only just been determined on.

What it evident from these reports is that there is never any reference to Fairey in the Gosport Aviation Co title. The Phoenix and Dick Kerr hulls have the Fairey name as their heading, *cf* Fairey/Dick Kerr, acknowledging Fairey as the designer, but not so for the Gosport Aviation Co. Therefore, even though it received a Fairey aerostructure, would it qualify as a Fairey design?

It was proposed that the Condor engines were to be mounted back to back with side radiators that yawed at 30-35° to the direction of flight. These radiators were for English Summer conditions. Auxiliary radiators arranged in series had to be allowed for in the final design.

In a project as large and geographically spread as this one often the small things were overlooked. In February 1919, Phoenix wrote requesting plans of the N-4 being constructed by Camper and Nicholson as they were to design the launching cradle for this boat. These cradles were essential for handling the boats out of the water.

Photographs of the construction of the Atlanta designed by Charles Nicholson accompany a series of articles on flying boat construction by David Nicholson in *Flight* during 1919. Nicholson commented that Charles Nicholson's *method omits the closely spaced transverse frames and beams; the transverse strength is obtained by fitting directly on the skin a series of strong web frames at widely spaced intervals.*[51]

The hull was shipped to Bradford where it was fitted with the aerostructure built by the Phoenix Dynamo Co. It was stored and eventually taken to Grain where the hull was subject to flotation tests. It appears that it never flew and was scrapped.

Fairey wrote in January 1919, that several of their draughtsmen did not return to work after the Christmas holiday. The company was *anxious to fill their places so as to expedite in every possible way the completion of the ... design.* They requested that a W.L. Morgan, an Acting Sergeant Pilot with No. 26 Training Depot Station and previously a Fairey employee, be released as they required his services as a draughtsman.[52] Other problems in additional to the usual design and construction problems arose and in the post-war climate, there appears to have been no rush to complete the machines. Fairey had to write to the Department in early August that the Phoenix Dynamo Co could not complete the centre section as they had been awaiting the correct drawings of this part for some three months. The Company had submitter them in July for approval and they appear to have got lost. At the end of August, it was proposed to mount a 75-mm Quick Firing gun in the Atalanta and Titania front cockpits.

A Minute of October 1922, recorded that experiments with very large aircraft had not been sufficiently successful to justify the construction of a 4,000-hp all metal flying boat "at present." There were two experiments in hand to develop large flying boats. One N.4 was erected at Grain but had not flown due to *difficulties that have been found.* The two principal ones were that water thrown up by the hull broke the propellers, and secondly, the failure of the servo-motor gear for operating the tail plane. The next attempt to fly the machine was stated to be at the end of November.[53] The other experiment was a hull, D of R Type 10, being built at Supermarine. This hull was an attempt to construct a hull light enough for use with a flying boat, but at the same time of suitable strength and shape to give good sea keeping qualities. Metal construction was being provided for in the estimates, being for a P.5 with steel hull. This latter was to emerge as the English Electric Kingston.[54]

The post-war financial position led to severe restrictions on the acquisition of flying boats for experimental purposes. The

Fairey N.4 Atalanta & Titania Specifications						
Source	**1.**	**2.**	**1.**	**3.2**	**4.**	**5.**
Type	**Atlanta**		**Titania**	**Atlanta**	**Titania?**	
Span upper	139 ft	140 ft	139 ft		140 ft	138 ft
Span lower						93 ft
Length	66 ft		66 ft		67 ft	65 ft 9in
Height (without trolley)		31				30 ft 6 in
Chord						12 ft 6 in
Mean gap						13 ft 0 in
Dihedral (lower wing)						2°
Wing area in sq ft	2,900		2,900			2,769
Weight in lbs						
Structural weight						
Empty				22,1781		
Useful load						
Crew				858		
Fuel gravity (200 galls)				1,460		
Fuel main fwd (400 galls)				2,920		
Oil (60½ galls)				581		
Deadweight				1,972		
Gross						32,500
Loaded weight	30,500		31,612	29,969		
Endurance (without military load)	7 to 9 hrs					8 to 10 hours.[3]

Notes: 1. Includes water, starting batteries & instruments.
2. Mean results of weighings on three occasions.
3. Without military load.

Sources:
1. Taylor, HA. *Fairey Aircraft since 1915*, Putnam, UK, 1984.
2. TNA AIR2/989
3. Isle of Grain Report N.M.270. Fairey "Atalanta" Flying Boat N.119. 4-Rolls Royce Condors. Weight and position of Centre of Gravity. 22.07.1922. Copy in AIR1/1204/204/5/2616.
4. "Fairey Seaplanes – Flying Boats and Aeroplanes". Advertising pamphlet. States machine has Linton Hope hull.
5. USN Report 28.11.1918.

Admiralty wanted a flying boat that could keep with the Fleet and operate from the open sea. In response to these concerns a conference was held on 29 November 1920, at which the members of the Air Ministry attended in an advisory role. To the statement that experiments should be conducted jointly by the Air Ministry and the Admiralty, the Air Staff replied that this *matter is continuously under consideration by the staff of the Director of Research, and experiments will be carried out as soon as the boats are ready.*[55]

Flying boat development was discussed at Coastal Area HQ in April 1923. This document noted that during the war the circular or Linton Hope type of hull was more efficient than the braced plank or "coffin" type hull, (that is the Porte type hull), for:-

(a). For landing and rising.
(b). Resistance in the air.
(c). Weight.

This is despite no Linton Hope hull being successful during the war.

At the Scillies during the war, with HMS *Ark Royal*, and in the Mediterranean, it had been proven that the mooring out of *small flying boats of the F.5 size* could be relied upon at the expense of the loss of performance in inverse proportion to the length of the cruise.

From the same sources it has been learnt that the F.5 sized boat is fairly satisfactory as a coastal patrol craft but its small size precludes its use for open sea operations, the sea keeping qualities improving with the size of the boat.

From (the) N.4 and N.3 types definite data has been obtained on the hull loading formula and it is now possible to plot a curve from the 100 Gnome F.B.A. boat to the 2,400 H.P. Fairey N.4., with sufficient accuracy to predict the water performance of any flying boat between these extremes.

The number of F.5 boats in service was stated to be 30 with an additional 39 F.2A boats that were, or could be, brought into service to replace a portion of the F.5 boats if required. Wastage was estimated at 33 $^{1}/_{3}$% per annum for crashes and an additional 10% for deterioration. 13 F.5 boats as an Initial Equipment would last the service 12 years if both types were used, or 5 years if only the F.5 was used.

It was believed that a number of experiments were being undertaken with a view to:-

1. Replace the F.5.
2. Produce an open sea cruiser flying boat.
3. Research work.

These were stated to be:

1. Further experiments with two N.4 type machines.
   This type have (sic) *great objections, both in design and construction, and apparently already taught all the lessons required, if considered in conjunction with the design of large land planes.*

   It is suggested that little useful purpose will be served by further expenditure on these craft.

2. N.3 Cromarty and Valentia – fitting of a new hull.
   These machines have already proven their worth and have given all the information that they ever will give. They have proven the unsuitability of consulta and coffin type hulls and have given valuable data on the water loading curve for boats of this size. It is thought that no further information will be obtained, by extending these experiments, that is not already known.[56]

The RAF wanted to build a 9 ton flying boat (the size of the N.4) along the lines of the Phoenix P.5 to replace the Felixstowe F.5 and was inclined to save funds by cancelling the experimental N.3 and N.4 types, keeping the F.5 and F.2A in service for "a slightly longer period" and by not purchasing an intermediate type, such a boat could be afforded and *brought into service by 1926 or 1927 at the latest.*[57]

In reply it was stated that not all the lessons that could be learned from the N.4 had been learned and that it was *hoped to get considerable information from taxying trials in the comparison tank and full size cleanliness of running etc.*

We have aerodynamic data from the V. Handley Page (the V/1500 four engine bomber) *but N.4 differs considerably in design from this and data obtained from flying tests should be of considerable assistance in the development of future large flying boats.*

As funding to get the N.4 into the air had already been incurred and it was recommended that this be allowed to proceed. *A contract is in hand to embody essential alterations in Titania which is fitted with flaps.* These "flaps" were the Fairey Patented Variable Camber Gear. A Fairey press release noted that the Titania was an improved type of the Atalanta adapted to give *extremely long range non-stop flights.*[58]

The change to metal hulls and post-war economy meant that the N.4 flying boats never had a chance to prove themselves. They were the only flying boats to carry the Fairey name.

Postscript:

In January 1919, C.R. Fairey wrote to the Sir Arthur Duckham of the Air Ministry that in *February last the writer had an interview with Lord Weir at which it was arranged that we should have a Contract for the design of the N.4 Flying Boat known as the "Fairey Atlanta", and further that we should supply drawings and assistance to Messrs. Dick Kerr of Preston and Messrs. Phoenix of Bradford in connection with the construction of these machines. We have written to Lord Weir asking him to confirm these facts to you.*

*Relying on the assurances given to us at these interviews we commenced work employing the whole of our Drawing Office Staff which we increased to enable us to cope with the work, set up tes*ting machinery, installed a representative at Messrs. Dick Kerr & Co. and generally proceeded with the design.

The writer goes on to describe that request for a Contract finally led to the Contract Department asking for the firm's terms on 8 June 1918. These were sent. In response to repeated letters the Contract Department finally replied rejecting the terms and stated it was now a matter for the Treasury Committee. Fairey then asked if they were in a position to negotiate the terms agreeing to modify the firm's proposal *as far as reasonable* to meet their views, but again had received no reply. On 19 October and 4 December Fairey had asked for a payment of £5,000 on account "as we were at considerable financial loss and the position was most unsatisfactory to us" but had received no reply to our request. Fairey concluded with a request for an early interview to settle this "long outstanding matter."

A Minute of 4 January 1919, noted that the *work is now nearly complete* and that the delay in concluding the matter, besides Fairey's financial position, *this delay is having a very bad effect on seaplane progress, which is even more important, from a commercial point of view, than aeroplane progress, as a glance at the map will show.*[59] It is not recorded when Fairey finally received his payment but it is assumed that this took place.

An Atalanta hull was said to have been used as a house-boat and was the subject of correspondence in *Air Pictorial* in 1957.[60] From the photograph it cannot be stated that the hull was indeed that of the Atlanta but it had "Atalanta" painted on the nose. The hull was at Felixstowe Ferry, near Felixstowe along with a Supermarine Southampton hull. In reply to the correspondence "Senrab" wrote that he saw N119 at Felixstowe in 1924 after it went to that station for open-sea tests. He identified the houseboat as the hull of the "Atlanta I". The Titania, N129, was flown at Felixstowe in 1925, and was *still airworthy when I saw it in 1928, but had apparently spent most of its time on the ground, and was no doubt disassembled soon afterwards, as it had disappeared when I visited the Schneider Trophy practice team in the spring of 1930.*[61]

It is thought that the Southampton hull is that that is now housed in the RAF Museum, Hendon. The fate of the other hull is unknown.

End Notes

1. Weekly Report of the Controller, Technical Department, 23.01.1918. In another Report mention is made of a May, Harding & May experimental hull for the Type N.3B. Fairey and Phoenix were reported as looking into the design of the aerial structure and would be forwarding proposals. The involvement of these three firms suggests that the N.3B was an outgrowth of the N.3 requirements but this author has not been able to find any other reference to the proposed type. (P.13 of Controller, Technical Department report. Loose page in RAFM J.M. Bruce Collection Box 64.)
2. Memorandum for Captain Schoefield, 28.11.1918; - pencilled on copy "For Commander Hunsaker." Copy in NAHC, Washington Navy Yard, DC.
3. Taylor gives the incorrect RAF specification XXII. Taylor, HA. *Fairey Aircraft since 1915,* Putnam, UK, 1974.
4. Memo: Lt Col J.G. Weir, Controller, Technical Dept to DGAP, *N.3 and N.4 Specifications,* 23.01.1918. The N.4 is referred to as an "Open Sea Machine." TNA MUN8/3.
5. Minute. DIS to D.1. 21.11.19. TNA MUN8/3.
6. Memo to DGAP: "Fairey Atalanta" signed by Lt Col Alec Ogilvie, Assistant Controller, for Controller Technical Department. 04.01.1919. TNA MUN8/3 Pt.2.
7. It was assumed that the two seaplanes referred to were the Atlanta and the Titania, however other documents suggest that the Titania was a later development.
8. Minute: From F.E.T. Hewett for Head of D.(S), *Recommendation for the Design and Construction of Hulls for N.4 Seaplanes.* 23.02.1918.
9. Minute: For information. 08.03.1918. TNA MUN8/3.
10. Minute: W. Weir to C.T.D. Commander Hewlett D.(S). 13.03.1918. TNA MUN8/3.
11. Memo to CTD through D, from Head of D(S). 25.03.1918. TNA MUN8/3 Pt.2.
12. Minutes of Meeting held on 17 April 1918. Subject: *Design & Construction of N.4 Experimental Seaplane.* TNA AIR2/989.
13. Minute: DIS to D.1. 21.11.1919. This Minute states that there is no record of "any order having been sent to Messrs. Fairey Aviation Co. for the design of these machines." TNA MUN8/3.
14. Letter. C.R. Fairey to Controller, Technical Dept. *RE: N.4 Flying Boat.* 15.05.1918. TNA MUN8/3.
15. Assistant Controller, Design to R.P. Wilson, ACS(S), *Re N.4 Experimental Flying Boat.* 18.05.1918. TNA MUN8/3.
16. Alec Ogilvie., A-CTD, Design to DGAP, 27.05.1918. TNA MUN8/3. It is interesting to note that Fairey was to receive payment for his ideas while those of John Porte, as a serving Naval officer, belonged to the Crown.
17. "Request for Requisition to build Experimental Aircraft for the General Benefit of Aeronautics." Form C. 06.05.1918. TNA AIR2/987. Six-Seater Seagoing Protected Bomber Flying Boat with 5 Eagle VIII Engines. Strength. Felixstowe Fury.
18. Memo: The development of the Large Flying Boat from the present type of Felixstowe Flying Boat. 08.03.1918. UK NA File AB/286/154/5 Part II.
19. According to Taylor, Fairey received a contract for two machines but due to shortage of floor space, they sub-contracted one machine to Dick Kerr and Co, of Lytham, St Annes. However, the Technical Committee Reports show that the contract for the complete seaplane was given to Dick Kerr.
20. Andrews gives the spelling as Fyffes on the Clyde. The author has not been able to find any information on this firm, assumed to be yacht builders.
21. According to Bruce Robertson May, Harding and May, had a boatyard at Hampton, Wick, Middlesex, and a premise on the South Coast. Contained in "Letter to the Editor", *Air Pictorial,* July, 1957, P.247.
22. A Memorandum of 17.10.1918, was apparently raised to obtain an official contract for Fairey. This stated that when the original design was prepared "insufficient data was available as to the practicability of employing low-lift wings with flaps, and owing, to the great weight per sq. ft. to be lifted at low speeds, high lift wings were adopted. This data has since come to hand and indicates that flap wings are perfectly feasible. Their employment will permit of an enhanced performance which may amount to a gain in speed of 10 knots." The <u>third or spare hull</u> at present being built was stated to be available for use with this aerostructure. (*Note to accompany request for requisition to build an experimental aircraft to meet R.A.F. Type XXXIII – Demand No. A.15.B.* Alec Ogilvie, Assistant Controller, Design.) The accompanying "Request for Requisition to build Experimental Aircraft to meet specific requirements" notes it was for "Design and construction of one machine. <u>Hull to be supplied to contractor</u>." (Author's underlining.) (Copies in TNA MUN8/3 Pt.2). Thus, it appears that a separate contract was raised to cover the machine that eventually was to emerge as the Fairey Titania.
23. A Report of 01.01.1919 states that the mock-up of seating and controls had been modified and "considerably improved." The tanks were sent to Dick Kerr. The tail unit had been detailed and a stress diagram was ready for checking. Inspector's Report of Fairey Aviation Co by Maj Bumpus, RAF. TNA MUN8/3.
24. The Phoenix Co became part of the English Electric Co in 1919. A photograph of the Atalanta built by English Electric Co appears on P.149 of *Flight* for 18.03.1924.
25. Phoenix Dynamo 4 Seat Flying Boat Twin Engine P.5

Cork. TNA AIR 2/58/AB286/154/5 Part II.

26. Report of Messrs Dick Kerr, Preston by Maj Bumpus on16.01.19. TNA 8/3.
27. Smith, R. *British Built Aircraft – Northern England, Scotland, Wales and Northern Island.* The History Press, UK, 2005. P.172.
28. Minute. From Wing Cdr Briggs to D of R, *N.4 Flying Boat, Report of visit to Lytham.* 24.10.1919. TNA MUN8/3.
29. Technical Dept, Aircraft Production, to Fairey Aviation Co Ltd for transmission to Phoenix and Dick Kerr, 06.11.1918. TNA TNA AIR 2/58/AB286/154/5 Part II.
30. "Flying Boat Hulls, P-5 and N-4 as designed by Lt Col Linton Hope, RAF." Copy in NAHC, Washington Navy Yard, DC. A slightly different report was printed as Technical Note No.17 by the Bureau of Aeronautics. The Report was made by a Mr S Truscott and the hulls were inspected at the May, Harding and May yards. Reprinted in *World War I Aero* No.146, P.70.
31. Memorandum for Captain Schoefield. "For Commodore Hunsaker" in pencil. 28.11.1918. The report appears to have been written by RH Griffin. Copy in NHHC, Washington Navy Yard, DC.
32. Referring to the hulls as the Dick Kerr hull only confuses the issue. The contracts were separate and it seems that even at this early date confusion was creeping into the story of the N.4. Nicolson, D. "Design and Construction of Flying Boats", *Flight*, 10.07.1919. P.919.
33. Nicolson, D. "Design and Construction of Flying Boats," *Flight,* 10.07.1919, P.916.
34. Report: *Comments on The Grain Test Report No. N.M.219B.* 17.10.1918. Comparing P.5 with F.5 (Nos. N.90 & N.86). UK NA File AB/286/154/5 Part II.
35. Taylor gives the period when the hull was transported by road to Lancashire as occurring in 1919.
36. A letter of 11.07.1919, from "D" to the D of R suggested that the Fairey Atalanta N.4 be tested at Lytham. *The only alternative would be to take the boat by sea round to Felixstowe, which would of course entail a number of Messrs. Dick Kerr's employees also going there.*
    I think that it should be remembered that this machine is far the largest that we have ever had to handle, and that the trials will probably take a very considerable time. TNA MUN8/3 Pt.2.
37. Isle of Grain Report N.M.270b. 15.08.1922. Copy in TNA AIR1/1204/204/5/2616.
38. DDR to D of R 30.10.1922. The Development of Large Seaplanes and Flying Boats. TNA AIR5/209/10/45/41.
39. *Flight*, 29.05.24. P.314.
40. Aeronautical Research Committee, *Forces and Moments (including those due to controls) on a model Fairey "N.4" Flying Boat "Atalanta", at various angles of yaw.* November 1924. TNA DSIR 23/1988.
41. Letter: Design to Fairey Aviation Co, RE*: "Fairey Atalanta" N.4.* 12.06.19. TNA MUN8/3.
42. The Fairey Aviation Co was notified on 22.11.1918 that the name of the N.4 would be the Fairey Titania. Lt Col AO (A Ogilvie?) Assistant Controller, Design, to Fairey Aviation Co. TNA MUN8/3 Pt.2. The name had been agreed to beforehand by Fairey and the Air Board.
43. Report on Visit to Fairey Aviation Co. on 5.11.19 by Sqn Ldr Miley, Head of DIS. 06.11.19. TNA MUN8/3.
44. "Atalanta's Story" in "Letters to the Editor" section*, Air Pictorial,* July 1957, P.247.
45. "Fairey Flying Boat", Report from *The Times* of March 4, 1921, reproduced in *Aerial Age*, Vol. 13, 1921. P.182.
46. This sequence of events is taken from Taylor. As the machine was contracted to Phoenix, it would appear that the hull would have come from that company and not Fairey at Hayes.
47. The company is still in existence building luxury yachts, but ceased its links with the original family in 2005. TOM Sopwith's steam yacht *Philante* was built by the firm.
48. Penrose (P.401) states that the Gosport hull was inspired by Porte whereas it is a Linton Hope derivative.
49. Robinson agrees that N118 was the Phoenix boat.
50. Maj Y.A. Bumpus Report of 01.01.1919, *Op Cit.*
51. Nicholson, D. "Flying Boat Construction," *Flight,* 24.03.1921. P.215. This article contains many photographs of the Atlanta hull under construction
52. Letter: Fairey Aviation Co, to The Controller, Technical Department, Aircraft Production. RE: *N.4. Boat.* 14.01.1919. TNA MUN8/3.
53. Minute. HM Cane(?), DDR (D) to D of R. 30.10.1922. TNA MUN8/3.
54. For the story of the P.5 see Owers, C.A. "The Phoenix P.5 Cork," *Cross & Cockade International*, Vol.45 No.1, 2014.
55. Naval and Air Staff remarks. Notes from meeting of 29.11.1920. TNA AIR5/209/10/45/41.
56. AVM A.V. Vyvyan, Air Officer Commanding, Coastal Area, to Air Ministry. 04.04.1923. TNA AIR5/209/10/45/41.
57. RAF to Air Ministry, 04.04.1923. The Development of Large Seaplanes and Flying Boats. TNA AIR5/209/10/45/41.
58. Copy with Fairey Aviation Co letter to The Controller, Technical Dept, 19.02.1919. TNA MUN8/3.
59. Memo to DGAP: "Fairey Atalanta" signed by Lt Col Alec Ogilvie, Assistant Controller, for Controller Technical Department. 04.01.1919. TNA MUN8/3 Pt.2.
60. "Atalantas Story" in "Letters to the Editor" section*, Air Pictorial,* July 1957, P.247.
61. *Ibid.*

Bibliography

Alexandrov, A. *Private Duty. Soviet Naval Aviation in the Baltic 1918–1925*. B& K, 2000.

Bell Davies, R. *Sailor in the Air,* Peter Davies, UK, 1967.

Brew, A. *Sunbeam Aero-Engines,* Airlife, UK, 1998.

Bruce, J.M. *British Aeroplanes 1914-1918,* Putnam, UK. 1957.

Bruce, J.M. "Fairey Campania", *Air Pictorial,* UK, May 1973. P. 142.

Dodson, C & Miller, J. *The day we almost bombed Moscow,* Hodder & Stoughton, UK, 1986.

Fairey Campania Type F.22 (275 H.P. Rolls-Royce) Rigging Notes.

Grey, C.G. (Ed). *Jane's All the World's Aircraft 1919.* David & Charles Reprints, UK, 1969.

Green, W & Swanborough, N. *The Complete Book of Fighters,* Greenwich Editions, UK, Revised Edition, 2004.

Fairey Seaplane Type III.B and III.C. Air Ministry, UK. 1921.

Flying Boat Hulls, P-5 and N-4 as designed by Lt Col Linton Hope, RAF. USN Technical Note. Copy in NAHC, Washington Navy Yard, DC.

Hagby, K. *Norwegian Civil Aircraft since 1919 an illustrated register survey.* Midland Counties Publication

Khairulin. M & Kondrastiev, V. *Aviation in the Civil War in Russia*, Tekhnika Molodeshi, Moscow. Translated by Alexandrov, A. Russia, 2000.

Khairulin. M. *Russian Aviation Colours 1909–1922, Vol.4 Against Soviets.* MMP Books, UK. 2018.

King, H.F. *Armament of British Aircraft 1909–1939,* Putnam, UK, 1971.

Kulikov, V. "Slavonic-British Air Corps over North Russia." *AviaMaster,* No.1, 1999.

Mason, F.K. *The British Bomber since 1914,* Putnam, UK, 1994.

Moss, P. *The British Civil Aircraft Registers 1919–1928,* Air Britain, UK, 1969

Mulder, R.J.M. *The Beginnings of Norway's Airlines, Part 1: 1918–1922,* European Airlines, Norway, 2018.

Nicolson, D. "Design and Construction of Flying Boats," *Flight,* 03.07.19. P.874.

Nicolson, D. "Flying Boat Construction," *Flight,* 24.03.21. P.213.

Nowarra, H. *Marine Aircraft of the 1914-1918 War,* Harleyford Publications, UK, 1966.

Penrose H. *British Aviation. The Great War and Armistice.* Putnam, UK, 1969.

Rennie, JD. "Some Notes on the Design, Construction and Operation of Flying Boats," from the *Journal of the RAeS*, 7th Meeting, 58th Session.

Robertson, B. *British Military Aircraft Serials 1911-1979,* Patrick Stevens Ltd, UK, 1979.

Smith, A. *The Man Who Built The Swordfish–The Life of Sir Richard Fairey*, I.B. Tarus, UK, 2018.

Snowdon Gamble, CF. *The Story of a North Sea Air Station,* Oxford University Press, 1928.

Sturtivant, R & Page, G. *Royal Navy Aircraft Serials and Units 1911-1919,* Air Britain Publication, UK, 1992.

Taylor, HA. *Fairey Aircraft since 1915,* Putnam, UK, 1974.

Technical Notes Fairey Seaplane Type III. RAF.

Types of British Seaplanes, Flying Boats and Ship's Aeroplanes. August 1918.

Woodman H. "The West Supports US," *Windsock*, Albatros Productions, UK, Vol. 14, No. 2, 1998. P.9.

Files in the RAF Museum J.M. Bruce Collection, UK National Archives & Smithsonian National Air & Space Museum, Washington, DC.

Afterword

Above: Fairey IIID N9777 bears a racing number on its rudder and fuselage. The aircraft is in the RAF post-WWI overall aluminium doped scheme. The IIID carried the Fairey company through the bad times of the 1920s.

In its 20 October 1921 issue, the Australian *Aircraft* magazine reported that Australia's first naval aircraft, "A.N.A.1", had been launched in the UK on the Hamble River, on 12 August by Dame Mary Hughes, the wife of the Australian Prime Minister. The aircraft was a Fairey IIID floatplane. Fairey's chief test pilot, Lt-Colonel Vincent Nicholl, DSO, DSC, took Mrs Coates, wife of the Commonwealth's Air Liaison Officer, General Seely and General Brancker, the Master-General of RAF personnel, on the official maiden flight after Mrs Hughes christened the aircraft "Mary" with the customary bottle of champagne over the propeller boss. The report continued that six IIID floatplanes were presently on board the SS *Boonah* en-route to Melbourne out of an order for 12.

Frank Barnwell, the designer of the Bristol Fighter, had joined the RAAF in Australia following the end of the war. He was in the UK in 1921 while serving in the RAAF and was asked to report on ship aeroplanes. He was taken to the Admiralty Drawing Office where he saw details of the Deck Landing Gear as fitted to HMS *Hermes,* and visited Parnall & Co of Bristol on 12 November where he saw H Bolas the designer of the Parnall Panther. *He gave me good information as to undercarriages, hook fittings, etc., for ship Aeroplanes.*

On 8th. November, I went down to Hamble with Colonel Nichol to the testing station of the Fairey Aviation Co. Ltd.

Mr C.R. Fairey had very kindly arranged that I should be allowed to fly a IIIC. machine belonging to his firm. This machine is practically identical with the III.D. machines recently sent to Australia, except that the rear petrol tank is removed and complete dual control is fitted.

Colonel Nichol took me out in the machine, gave me a passenger flight and then let me fly the machine for about 40 minutes, during which time I took-off and landed the machine 6 times.

The machine struck me as being noticeably easy to fly, to take-off and land. The few points perhaps worth noting are:

1. *In taking-off it appeared necessary to hold the machine down gently until planing, then to ease back the control column gently as the speed increased until the machine flew itself off.*
2. *As the engine is set at an angle to the fore and aft centre line, the machine required R.H. rudder with the engine full on climbing,*

L.H. rudder with engine throttled right down gliding; but, for the same reason the rudder is not heavy under either condition

3. *There appeared to be a slight tendency to overbank on turns, as opposed to the behaviour of most land machines.*[1]

What ever the reason, probably the lack of funds, Australia did not go ahead with purchasing any ship aeroplanes, and the Navy lost their Fairey IIID floatplanes to the RAAF. In Australian service the IIID gave sterling service as did the IIID floatplanes in Portuguese and Netherlands service. They were continually improved and developed into the Fairey IIIF that served from ships and land in Argentina, Chile, Greece, New Zealand as well with 14 RAF squadrons and five FAA squadrons.

From a hastily modification in order to provide a machine to serve in Russia, the Fairey III series cemented Fairey as a manufacture of aircraft. It is perhaps ironic that the most famous aircraft bult by Fairey in WWII was the obsolete Swordfish. Post-WWII Fairey diversified and continued to build aircraft up to the successful Firefly carrier born reconnaissance-fighter and Gannet anti-submarine aircraft. The company's venture into helicopters was not profitable. In 1977 the Fairey Group went into receivership and was nationalised by the Government.

Fairy aircraft served successfully during WWI with little fanfare. In WWII their name was indelibly attached to the sinking of the *Bismarck*. Post-WWII the Fairey Delta 2 took the world air speed record for the UK. As with so many of the British companies that were started by the early pioneers, it fell into bad times and was nationalised in the effort of the British Government to rationalise an industry that had grown unable to support the pioneer companies.

End Note

1. Memorandum "Further Notes on Deck Landing Gear on Seaplanes." By F.S. Barnwell, Sqn Ldr RAAF. Sent to Australia 13.12.1921. Copy in Australian National Archives. CRS A705/1 Item 199/17. "Handling Flying Boats and Seaplanes."

Above: The catapult on HMS *Slinger.*

Made in United States
Orlando, FL
04 November 2026

90205911R00136